OBJECT LESSONS

A book series about the hidden lives of ordinary things.

Series Editors:

Ian Bogost and Christopher Schaberg

Advisory Board:

Sara Ahmed, Jane Bennett, Johanna Drucker, Raiford Guins, Graham Harman, renée hoogland, Pam Houston, Eileen Joy, Douglas Kahn, Daniel Miller, Esther Milne, Timothy Morton, Nigel Thrift, Kathleen Stewart, Rob Walker, Michele White.

In association with

BOOKS IN THE SERIES

phone booth

ARIANA KELLY

Bloomsbury Academic
An imprint of Bloomsbury Publishing Inc

B L O O M S B U R Y
NEW YORK · LONDON · OXFORD · NEW DELHI · SYDNEY

Bloomsbury Academic

An imprint of Bloomsbury Publishing Inc

1385 Broadway
New York
NY 10018
USA

50 Bedford Square
London
WC1B 3DP
UK

www.bloomsbury.com

**BLOOMSBURY and the Diana logo are trademarks of
Bloomsbury Publishing Plc**

First published 2015

Library of Congress Cataloging-in-Publication Data
Kelly, Ariana.
Phone booth / Ariana Kelly.
pages cm. — (Object lessons)
Includes bibliographical references and index.
ISBN 978-1-62892-410-7 (hardback : alk. paper) — ISBN 978-1-62892-409-1
(pbk. : alk. paper) 1. Social interaction. 2. Telephone booths—Social aspects.
3. Telephone calls—Psychological aspects. 4. Telephone—Social aspects. I. Title.
HM1111.K45 2015
302—dc23
2015010323

ISBN: PB: 978-1-6289-2409-1
ePub: 978-1-6289-2411-4
ePDF: 978-1-6289-2412-1

Series: Object Lessons

Typeset by Deanta Global Publishing Services, Chennai, India
Printed and bound in the United States of America

CONTENTS

1 DISCONNECTED

Standing in a garden in Ōtsuchi, a small town in the Iwate prefecture, on the east coast of Japan, there is a nonworking telephone booth that has nevertheless been used by more than ten thousand people since the spring of 2011, when a 9.0 earthquake and massive tsunami killed fifteen thousand people and dislocated hundreds of thousands more. Built by Itaru Sasaki, a sixty-nine-year-old resident of the area, so that people could communicate with loved ones who were killed or missing, the wooden-framed booth—which has plate glass windows and a door that closes—is named "Kaze no Denwa Box," or Phone Booth of the Winds. Although he installed a rotary dial telephone within the booth, Sasaki never connected the line; instead, there is a small notepad on the shelf beneath the telephone where people can leave messages and trust that the wind will carry the contents to their intended recipients.

Ōtsuchi was one of the areas most devastated by the tragedy. The town lost 10 percent of its population, most of its six hundred and fifty fishing boats and all of its sea farm business. The waves that poured over the reinforced-concrete sea wall were nearly thirty-feet tall, rising level with the clock face on the second floor of Ōtsuchi's town hall building. While the frame of the building remains, the town records are gone, swept out to sea, and that is the case for much of the area—what abides is the carapace of what was. After the last waves of the tsunami receded into the ocean, a ferry perched precariously atop a two-story building. Fires that began in the wake of the disaster burned for days because the fire department had been totally destroyed. Even before the earthquake and tsunami, though, people were worried about the future of Ōtsuchi. The fishing industry was diminishing, as was the job market in general. Now, younger families are understandably leery of building again so close to the coast.

In such a stricken environment, it might have been easy to perceive this disconnected phone booth as a whimsical art project incommensurate with the scale of loss experienced by the survivors for whom it was intended. But quite the opposite has happened, and for the past several years there has been a steady stream of visitors to the booth, both from Ōtsuchi and other parts of Japan. Perhaps it provides a necessary terminus, a destination in a place marked by eradication. Perhaps one antidote to tragedy is useless beauty, or just uselessness. In early January of 2015, almost four years after the tsunami, when high winds knocked the booth over, shattering its windows and ripping off its door, money poured in from all over Japan for its repair.[1]

2 HERMIT'S HUT

I keep thinking about a famous writing exercise someone once mentioned to me in passing, but which has stayed with me ever since. The exercise entails describing a barn from the perspective of a man whose son has been killed in the war, without ever mentioning the son, the death, or the war. I've never completed the prompt, and yet I think about this barn on a regular basis.

It's connected, in my mind, to an empty phone booth I once saw nearly every day for the better part of an autumn and early winter when I was living in Ireland. It was 1999, and I was renting a room in a tourist town on the southwest coast, in County Kerry, waitressing in the purportedly high-end restaurant of a mediocre hotel. The phone booth stood in an empty field on the opposite side of the street from the hotel. It was the rainy season, and I'd been traveling around the northern part of the country for a couple of months without contacting anyone. When my temporary work permit came through, though, I settled in the first place that gave me a job. Those were still the early days of the cell phone, and I didn't have one, nor did I want one.

I was glad to be out of touch, was in fact hoping to disappear for a while.

Almost all of the people who worked at the hotel were student travelers from Europe trying to improve their English, or, like me, happy to be in an English-speaking country. But the hotel itself was run by a family who clearly thought of themselves as Irish aristocracy. Maybe they were, but the cracks and stains that marred all but the most obvious surfaces did not bode well for their legacy. We were being overworked and underpaid, of course, which is why all of us got jobs there so easily, and why no one felt any loyalty to the establishment, such as it was—people came and went all the time in a constant stream of Anglo-European accents that transformed the atmosphere into something foreign and exotic.

The work was relentless: six hour shifts running back and forth between a dining hall that looked longingly onto the rainy Irish countryside through a bank of floor-to-ceiling windows, and a windowless industrial complex buried down a flight of stairs that sufficed for a kitchen. But time passed quickly, almost mystically, and there were long periods of unconsciousness, when my body seemed to move of its own accord, as efficiently as a good snowstorm dispenses with the niceties of autumn. Communication among the staff was distilled to its utilitarian essence, until one word seemed sufficient to stand for a meal, a day, whole nations, and belief systems. Linguistic flourishes were reserved for the string of profanities loosed if something was lost in translation. Everything else—the clamor of the customers, the anger of the cooks, even the thunder of an evening storm—was white noise.

After such din, the booth was particularly appealing. It seemed melancholy, adrift in the middle of the field, as if it had absorbed all the rain that had ever fallen and ever would fall, as it had absorbed all the conversations that had and would come to pass by virtue of its presence. Not that I ever saw anyone using the phone; already on the cusp of obsolescence, its existence went unacknowledged by tourists and locals alike. Solidly placed, sturdily built, it was anachronistic even then, suggesting as it did a distinct point in a field whose point wasn't clear. But the rain that fell during one week in broad slanting strokes continued to fall during the next, and at night the word TELEPHONE inscribed in dark letters along the white lintel glowed alluringly. "When we are lost in darkness and see a distant glimmer of light," writes Gaston Bachelard, "who does not dream of a thatched cottage, or, to go more deeply into legend, of a hermit's hut?"[1] I crossed the street one early evening after my shift ended, and instead of biking home in the downpour, I stepped into the booth—my hermit's hut—closed the door, and listened to the hushed murmuring of the atmosphere. The next morning the air would be thick with the smell of burning peat, and I would wake up at dawn to take the early shift. But that evening I made the first of many phone calls to the few people I was interested in speaking with at the time, cradling the phone against my shoulder, leaning against the trapezoidal counter, twirling the cord absentmindedly in my hand.

Now, so many years later, what comes back to me is not so much the particular details of those conversations, but their prevailing moods: urgent, intense, acute, also sodden, amorphous, lonely—time measured out in five- and ten-pence coins although it was really being doled out in the denominations of five by which I could add credit to my calling card. Communication had itself achieved a meter, the meter of crisis, and there is no more appropriate technology in a crisis, existential or otherwise, than a phone booth. Movies and books taught me this, but so did the object itself. The space was at once intimate and anonymous, private and public. Within it my dramas were unique and universal, and my emergencies took their place in a long lineage of emergencies.

What John Gardner implies in his exercise about the barn is what the poet James Merrill explicitly says when he advises writers, "You hardly ever need to *state* your feelings. The point is to feel and keep the eyes open. Then what you feel is expressed, is mimed back at you by the scene."[2] The point is clear: the external world is inflected and informed by the internal, if we can make ourselves pay attention enough to notice. One only embroiders it to observe that phone booths were places where we often *did* state our feelings, and arguably more openly than elsewhere.

Rain ran down the glass and resounded on the metal roof. Cyclists floated by in reflective gear like spectral fish in deep water. The field edged up against the borders of a national park, where some of Ireland's most imposing mountains loomed invisibly in the darkness. I was grateful for the shelter.

3 OUR SPEED

Calls made from pay phones and phone booths create their own mode of communication. It's a mode that revolves around urgent contact, finite contact, emergency contact, ubiquitous contact, democratic contact—the phone booth could provide asylum to anyone, *be he ne'er so vile*. It's a mode that wears its heart on its sleeve, so many public phones being etched with the tragedies of their users that their tragedies become shared tragedies. It's a mode that offers the promise of privacy and the possibility of anonymity. It's a paranoid mode. It's a mode, once upon a time, that seemed *more our speed*. It's a mode that is in motion and also a pause in motion. As Holden Caulfield wanders through New York City, delaying his arrival home after being kicked out of prep school, he is constantly entering and exiting phone booths, never quite knowing what he wants to say when he gets someone on the line, and often just sitting, as if phone booths are places to simply *be*, as much as they are places from which to call.

One can parse this distinction still further to distinguish between the calls made from the protective enclosure of a phone booth and those made from the more exposed kiosk.

To speak on a telephone in the middle of an empty Irish field or on the corner of a street with nothing but the broad sky as one's roof is a far different experience than speaking in these areas from within an enclosure. An enclosure carries with it a set of expectations—of safety and shelter primarily, but also of civility, order, a distinction being made between the natural and human world. As Gaston Bachelard so memorably articulated, a lamp resting in the window of a house on a dark night draws its meaning from casting its glow both inward and outward, creating warmth within and illuminating the darkness without. This is why one of the most disturbing (or comedic, depending on your mood) scenes in the 1988 remake of *The Blob* is when a woman takes shelter from the amorphous monster in a glass phone booth, only to find that the blob can seep through the seams of the

FIGURE 1 *The Blob*, dir. Chuck Russell, 1988.

box. Even though she succeeds in making a call to the town's sheriff, it's too late.

If the structure of the booth offered a degree of sanctuary and privacy in addition to the convenience of the phone itself, then the stand-alone kiosk focused only on convenience, its primary virtues as a structure being that it was a space-and-time-saver. These virtues are familiar virtues, touted as much or more today than they were when public phones were in the prime of their use. What have we done with all of the space and time saved? Ad copy portrays kiosks as perfect for the modern age—easily cleaned, used, and multiplied. On a street or in a building, a kiosk does not set itself apart from its environment, but rather becomes an element of the fray, a key accoutrement for a life spent in perpetual motion. One can use it, theoretically at least, without breaking one's flow through space and time. In its strict utilitarianism, though, the kiosk is a microcosm of a society that has come to prize speed and efficiency more than atmosphere and comfort. In the 1983 film *Trading Places*, when Paul Gleeson's Clarence Beeks is imparting some crucial information about crop estimates to the Duke brothers from a pay phone in Washington, D.C., a woman appears who wants to use the phone; Beeks tells the brothers to "Hold on," turns to the woman, tells her to "Fuck off," then returns to his conversation. It's a kiosk rather than a phone booth, though; he's standing in the rain with an umbrella.[1]

But this characterization of the kiosk is perhaps unfair, given that the enclosure of the booth was often abused, making crimes possible that wouldn't have been possible

FIGURE 2 *Trading Places*, dir. John Landis, 1983.

without the concealment offered by the structure. People have been raped, murdered, and mugged in phone booths. Drug dealers have used them as drop-off points. Terrorists have used them as redoubts. The general public has used them as places to piss, screw, and shit. A disturbing number have been the sites of suicides. Phone booths, unlike hotel rooms, are not cleaned after they are used. The slate is not erased, and so it accretes, telephone call by telephone call: the metonymic associations of its users.

With or without an enclosure, however, pay phones were created to assist people in difficult situations. Ask anyone now to describe the last time they used a public phone, and most will invoke a national, environmental, or personal disaster. Some will describe the morning of September 11, when the cell network went down in New York, and cell networks weren't fully national anyway, so using a cell phone out of state would result in massive roaming charges. Others will recount the days immediately following Hurricane Sandy, when there were lines of people waiting to use a working pay phone near St. Mark's and an entrepreneurial man was making a cool profit by charging two dollars for a dollar in change. In an essay from 2003, Kathryn Chetkovich describes being at a writing colony and making daily ten-minute calls to her ailing eighty-one-year-old father from a phone booth through whose small window she catches glimpses of the fellow writer she is falling in love with at the time. When Edward Norton's apartment blows up in David Fincher's 1999 *Fight Club*, he calls Tyler Durden (Brad Pitt) from a pay phone to see whether he can stay with him. We later learn, of course, that Tyler is merely an aspect of Norton's personality—he was calling himself.

In her 1970 novel *Play It As It Lays*, Joan Didion's main character Maria Wyeth senses an atmosphere fraying at its edges when she observes an unnaturally long line to use the pay phone at the grocery store: "The telephone in the apartment was out of order and she had to report it. The line at the pay phones in Ralph's Market suddenly suggested to Maria a disorganization so general that the norm was to have either a disconnected telephone or some clandestine business to conduct, some extramarital error."[2] As ubiquitous as elevators, but used much more rarely, public phones are intimately linked with varying degrees of disorder—from the incidental to the truly catastrophic, which is the context that gave rise to their existence in the first place.

In 1888 William Gray, the son of Scottish immigrants, was frantically combing the streets of Hartford, Connecticut, to find a way to call a doctor for his sick wife. Hartford was then one of the richest cities in the United States, made wealthy by manufacturing and later by insurance. Just twelve years after Alexander Graham Bell patented the telephone, the sky was virtually opaque with a complicated web of phone lines (each connection requiring its own line to a central switchboard), but there were few telephones available for public use. Those that were accessible existed in secluded areas of public spaces such as train stations and pharmacies and were regulated by people (usually women) who worked for the telephone company and accepted money for the use of the phone. Before that, many businesses had simply offered their phones for public use, free of charge, as a public service.

When he finally located one in a nearby factory, Gray was at first denied the use of the phone by a company official who said the phone was only for customers. By the time he convinced the official that his situation was serious, a business opportunity was beginning to stir in the mind of a man who was a machine polisher by trade, and who had already invented a well-received chest protector for baseball players. In less than a year, Gray had patented the first coin-operated pay phone, having entrusted George A. Long, a sixteen-year-old apprentice at the Pratt and Whitney factory, to engineer a working machine based upon his sketches. The design went through several iterations before it became more or less what we have today: an automated, prepay machine.[3]

Phone booths were centers of community in rural areas and necessary spaces of privacy in urban areas. They were the precursors to the ubiquitous connection we have today, and an early stage of the twenty-four-hour news cycle we now inhabit. When, in 1921, Jack Dempsey defeated the French war hero Georges Carpentier in four rounds in a hastily constructed arena in New Jersey, numerous spectators and listeners swarmed all available telephone booths to spread the news of the historic boxing match as quickly as possible: "There wasn't a telephone booth within blocks of the Square which didn't have a line of waiting men and women in front of it, waiting to telephone news of the fight to their friends."[4] More importantly, phone booths often dramatized human need, which is what they ended up fulfilling. One of the most crucial scenes in Roman Polanski's *Rosemary's Baby* is when Rosemary, convinced that her husband and neighbors are colluding against her unborn baby, escapes her apartment in order to call a doctor from a phone booth. In a four-minute shot, we inhabit the claustrophobic space of the booth, which is an apt representation of Rosemary's psyche, diminished as it is by fear and paranoia. From her perspective, every aspect of the world outside of the phone booth—even the man standing on the street, waiting to use the phone—is inflected with the sinister.

Yet few pieces of technology have suffered a more graceless end than pay phones and the booths that once enclosed them, most being systematically stripped of anything remotely valuable until, by the end, there is only a metal husk, a few stray wires twisting unnaturally upward, as if still aspiring to make something work. If at one point these structures that were specifically designed for a solitary occupant offered us a kind of mirror, then by now what they provide is mostly surfaces, often evocatively distressed, whose communicative action occurs in how people choose to deface them or in what products they advertise.

FIGURE 3 Phone kiosk. Photo © David Eng, used with permission.

Of course, pay phones still receive a significant amount of use, particularly by immigrant communities for whom pay phone calls can be considerably cheaper than monthly cell service, and in rural areas where cell phone service tends to be sparse and inconstant. But pay phones and landlines are largely acknowledged to be obsolescent, their last moments of glory manifesting during natural and manmade disasters and in shows like *The Wire*, which is premised on pay phones and pagers being used for drug deals because cell phones can be tapped too easily, as well as the podcast *Serial*, which revisits the 1999 murder of a high school senior in Maryland, a significant chunk of which hinges on a call being made from a phone booth that some people can remember and others cannot, and whose existence remains unproven.

Although the majority of phone booths come to rest in scrapyards, some have been resurrected, repurposed by needs as much spiritual and aesthetic as practical. Over the past fifteen years phone booths have been whimsically and practically transformed into aquariums, fountains, homeless shelters, lending libraries, galleries, and showers. In Kansas, a number of defunct phone kiosks have become "prayer booths," where one can kneel down on a padded platform, place oneself within the booth or kiosk so as "not to impede others," issue a prayer and then, before leaving, return the kneel bar to its upright position. More prosaically, some phone booths have been transformed into miniature medical centers, containing defibrillators in lieu of telephones, asserting the persistent usefulness, rather than the poetry, of these spaces.

But by far the greatest numbers have been turned into free Wi-Fi hotspots, the younger generation of technology seeking to reconcile with the older. The current effort to recycle and reuse, repurpose and otherwise resurrect an entity that has served as the backdrop for countless dramas suggests how emotionally attached we remain to an object that once provided for us so well, and perhaps to the world that provided us the object. "The past is hidden somewhere outside the realm," writes Marcel Proust, "beyond the reach of intellect, in some material object (in the sensation which that material object will give us) of which we have no inkling."[5]

The phone booth suggests a world in which communication is precious, urgent, clandestine, contingent. People talking in phone booths and at phone kiosks often exhibit an angle of repose—what is, in geology, "the maximum slope, measured in degrees from the horizontal, at which loose solid material will remain in place without sliding."[6] Virtually no one stands up straight; nearly everyone leans, usually against the triangular shelf beneath the telephone, or against the wall of the booth itself. They follow the line of the telephone, which follows the line of the face. It is an angle of nonchalance, absorption, self-importance, seduction; they are on the verge of sliding. David Bowie played with this lean beautifully on the back cover of his album *Ziggy Stardust*. In the photograph Bowie, as the alien Ziggy in a turquoise jumpsuit, stands in a British telephone box, arm resting on his hip, the other on the telephone, ready to call home from London.

4 THE PHANTOM PHONE BOOTH

Phone booths have proven themselves infinitely malleable, easily manipulated into other objects. It seems perfectly natural to turn a phone booth into an aquarium, for example, because, like the phone booth, an aquarium stresses the ideas of encasement and microenvironments, of inhabiting one world while being observed—or observing—another. For similar reasons, phone booths make surprisingly effective galleries and libraries, containing as well as displaying their contents, whether it is an individual or a painting or an assortment of books.

So what makes a phone booth a phone booth? Although the space has been reimagined many times, the basic booth is quite simple: a rectangular enclosure, roughly the size of a coffin or a confessional, with three walls, a roof, and a telephone. Anything else, including the door, light, desk and even phone book, is an optional feature that serves as a flourish on the essential design. Pay phones assume various shapes in other countries, but in America they are generally rectangular, with a coin depository at the top beside a lever that can be turned to retrieve one's change and a place to collect excess change at the bottom, often used as a repository for anything but coins. Next to the change collector is a coin vault, tantalizingly indicated by a keyhole and subject to more abuse than any other part of the phone. (A story from the 1920s reported a woman stealing more than three thousand dollars in change from pay phones, nickels, dimes, and quarters spread across her apartment floor like treasure in a dragon's lair.)

The touchtone dial in the center of the phone is appealingly tactile—a kind of braille in a world becoming less and less material—and recalls the control boards of older computers and calculators. The sound each key makes is also atavistic, an auditory affirmation of connection being made, lines being strung. One of the central ironies of making a call in a phone booth and, more to the point, of being in a phone booth, is how isolating an experience it can be. This is why the operator, who is assessable from every phone, is more of a presence in a phone booth than anywhere else: mediator, intermediary, facilitator. A thin band running along the top of the phone booth was the perfect place to display the corporate logo of the phone booth, and many kiosks and phone booths have gone through successive rebranding as telephone companies have perpetually reconfigured themselves throughout the twentieth and twenty-first centuries.

It's not uncommon to see a restored wooden phone booth in an urban restaurant, but most modern phone booths are made of the more durable aluminum—and are not there to create a vintage atmosphere—with a phone generally placed at eye level, above a shelf that sometimes holds a phone book connected to the booth by a steel chain or clip, and often encased in plastic. Not that these precautions did much good—most public phone books are weathered beyond repair or are missing significant sections, usually just the section one needs. This chain and encasement say something important about how vulnerable the booth was to being misused and vandalized, but also, as a structure, how worldly, how intimately bound to the fortunes of the people who use it.

Phone booths could exist in groups—in the flank at Grand Central Station, for example, or the fleet of wooden booths that still stands in Philippe's one of Southern California's landmarks—but they most often existed and exist alone, interrupting space at regular intervals. Some phone booths have seats, but most don't, putting a necessary cap on conversation and highlighting that this is a temporary space, made for temporary occupation. The phone booth is always pushing people onward. In 1936, when a woman at a Hartford bus terminal spent more than four hours on the phone in a phone booth, the terminal authorities alerted the police.

The public phone booth calls attention to itself by its signage, and indeed, there is something both inviting and anticipatory about phone booths. It is a public space that is truly public—available twenty-four hours a day, to anyone who wants to use it. They stand with three walls or an unlocked door, waiting to be used. Unfortunately, today they tend to be inviting only from a distance. Close up they are lacquered with secretions, graffiti, physical abuse. It is difficult for the most nostalgic among us to wring romance from the phone booths and kiosks that remain, and yet, to be sure, some do. Whereas once phone booths were so common as to be invisible, now their obsolescence renders them curiously conspicuous, although often not recognizable to those younger than, say, thirty. When the *New York Times* ran a story about a phone booth recently, they coyly explained what a phone booth was.

While the modern phone booth anticipates being abused and violated, the experience that most closely compares to the phone booth is that of the psychiatrist seeing a succession of patients, because no matter what the contents of the conversation, from the banal to the catastrophic, people are revealing themselves. "Times in a marriage we went there," writes the poet Brenda Hillman, "to complain or flirt. . . . Let us mourn secrets told to/Fake wood and the trapezoidal seat."[1] This most humble and basic of structures—the lean-to of communications—has been privy to the keenest expressions of grief, envy, jealousy, love, joy, shame, and anxiety. It is a structure that encouraged truth and opposed obfuscation, but of course, phone booths could precipitate psychological crises as readily as they helped people work through them. In Thomas Pynchon's *Bleeding Edge*, which takes place in New York City in the aftermath of September 11, the malfunctioning phone booths are symptomatic of a larger disorder and where a therapist could readily find potential clients:

> One day Shawn was in a phone booth here in town, out on the street, one of those calls he really needed to make, everything possible was going wrong, he kept shovelling quarters, no dial tone, robots giving him shit, finally working himself up into the usual NYC rage, slamming the receiver against the unit while screaming *fucking Giuliani*, when he heard this voice, human, real, calm. "Having a little trouble, there?" Later on of course Leopoldo copped to drumming up business this way, hanging around places where mental-health crises are likely to occur, like NYC phone booths, after first removing any out-of-order signs.[2]

5 SAY ANYTHING

In 1961 the New York Telephone Company installed fifty glass and gold-tinted aluminum booths off of Fifth, Park, and Madison Avenues, partly to construct public spaces that were more "in keeping with the tone of the neighborhood" and partly to encourage people to have longer, more involved conversations rather than the hurried, impulsive exchanges that usually occur on public phones. They were hoping to rearrange the geography of telephoning, where the most profound communication takes place in the privacy of one's home. Around the same time the American Telephone and Telegraph Company (AT&T) announced that it would have "glass, semicircular and serpentine-shaped booths"[1] at the upcoming New York 1964 World's Fair, which was dedicated to "Man's Experience on a Shrinking Globe in an Expanding Universe," and dominated by the space-age aspirations of the time. The booths would contain two of the newest innovations in telephones: touchtone dials and picture phones, distant precursors of Skype and FaceTime, thus emphasizing the booth's validity as both architecture and technology.

Gold booths might seem extreme (the *New York Times* wrote a scathing critique of the endeavor: "It is not the booths we object to; it is the gold."[2]), but the New York Telephone Company was really only trying to imbue phone booths with the glamor and luxury that characterized the first prototypes. Interstitial spaces, neither wholly public nor wholly private, provisional yet permanent, the earliest phone booths were nevertheless lovely places to be. The liminality of the spaces was not reason to give them short shrift. "We offer telephone subscribers a soundproof folding Telephone Booth at a moderate cost," states an 1892 brochure from the American Telephone Booth Company. "These booths are made double, with an air space between the inside and outside case. This renders them sound proof, insuring absolute privacy when using the telephone. They are made of either oak or cherry—with plate glass windows, and are highly finished—thus adding an ornamental article of furniture in any office."[3] Early phone booths, like the first movie theaters and elevators, were acutely beautiful examples of craftsmanship: wooden, spacious, inset with carpet and cushions, as if to suggest, subliminally at least, the value we once placed on transitional spaces. These areas, which were neither here nor there, were in fact designed to help people get from here to there and readily lent themselves to moments when private life spilled into the public. A phone booth was not just a convenient place to have a conversation—it was a significant place.

Long before the public telephone, and the telephone itself, existed as public commodities, Thomas Watson, Alexander Graham Bell's young assistant, improvised the first booth by creating a tunnel and enclosure with blankets and a barrel hoop in his apartment in Boston, where he shouted down the line to Bell, who was on the receiving end in New York.[4] It was the greatest distance they had tried to overcome and Watson's voice was compensating for the persistent weaknesses of the machine, much to the dismay of the woman who rented them the rooms.

In 1876, when Bell patented the telephone, the world was comparatively quiet. Only one outdoor telephone line existed in the world, an iron wire strung between Williams's shop, where Watson initially worked as an engineer, and Exeter Place laboratory, where Bell and Watson conducted many of their experiments. Trolley car and electric light systems did not yet exist; radio did not yet exist. Nevertheless, Watson and Bell soon discovered that the best times to test a new phone were nights and Sundays, because during the daytime the "quietest city room during the day is vibrating with a complex of sounds blended into a loud hum to which we are so accustomed that we don't ordinarily notice it."[5] Still, it was quiet enough that sometimes late at night Watson could hear "stray electric currents" that were "mystic enough" to suggest to him that the sound originated "from explosions on the sun or . . . were signals from another planet."[6] A devotee of Spiritualism, Watson found a kinship between the spirits invoked during séances around the parlor table and the electrical currents he and Bell manipulated to transport sound. Watson was not alone in his mysticism. In 1902 Guglielmo Marconi transmitted the first radio message, an accomplishment that led him to believe technology would become sophisticated enough to retrieve the sounds of Jesus delivering his Sermon on the Mount.[7] Although Watson later amended his theory, averring that the sounds were probably static currents, not signals from another planet, what he emphasizes is that a few years later these sounds were inaudible anyway, drowned out by ambient noise.

A few years after Watson's makeshift booth, in 1883, the first formal patent was filed for a phone booth with screened windows, a ventilator, and a small desk. It was even mounted on wheels, so that a customer could potentially move it to a quieter space. These early booths, considerably wider than the sleeker models that eventually replaced them, were installed in the lobbies of hotels and rail stations, as well as in restaurants and pharmacies. Often referred to as "cabinets" rather than booths, the telephone was something to be stored, to be put away and taken out, not perpetually present. It was a chamber that protected the human voice, the faculty Bell had taught to Watson to love, and defended if not created a quiet that no longer existed naturally.

For traveling salesmen and other insolvent entrepreneurs, telephone booths in the lobbies of public buildings were the only affordable places to do business. For these itinerant people, described by A. J. Liebling as the "Telephone Booth Indians," the phone booth was of material and emotional relevance, providing "sustenance as well as shelter, as the buffalo did for the Arapahoe and Sioux."[8] In 1929, the mayor of Philadelphia removed all phone booths from the city hall because too many employees were using them to place bets on horse races. This offense wouldn't have been as egregious, perhaps, if the men had been winning more than they were losing, but, according to a short article in the *New York Times*, "wives of a number of City Hall workers had declared that their husbands were losing large portions of their pay due to their inability to pick winners at various tracks across the country."[9]

While indoor booths flourished immediately after their invention, the first American outdoor booth did not appear until 1905, on a Cincinnati street. The outdoor booths were initially also made of wood, although they weren't outfitted as luxuriously as many indoor booths. Like the telephones of the late Victorian and Edwardian eras from which they originated, these ornate wooden booths not only provided privacy; they also suggested the telephone was more of an aesthetic object than a machine. Outdoor booths had to make concessions to the environments, however, and were necessarily more utilitarian. Even with the privacy afforded by a wooden booth, however, talking on the phone outside remained an uncomfortable concept for many, and the outdoor booth did not become popular and widespread until the 1930s. By 1949, the most popular telephone booth manufacturer, in Queensboro, New York, was slated to build 4,150.[10]

The telephone box

The British understood the importance of the space so well that they organized a contest under the auspices of the Royal Fine Arts Commission to garner a compelling design. The United Kingdom Postal Service, which at the time administered both telegraphs and telephones, had released a prototype in 1920, but the model never became popular in any but low-income areas. In contrast, the design that won—which was drawn by Sir Giles Gilbert Scott, the architect also responsible for the building that contains the Tate Modern and Liverpool's Anglican Cathedral—seemed to embody some of the imperious dignity of the British Empire itself. Standing like a classical column on a raised plinth, with a domed roof that, after 1926, bore an insignia of the British crown above the door, Scott's booth became a national symbol in the way American phone booths never did.[11] Constructed from "steel and small pane glass," the sturdiness of the booth not only responded to the vagaries of British weather but the recent memory of the German air raids on England, which had ended a mere six years before Giles began his design.[12]

Since 1924 Scott's design has gone through a series of modifications. After 1953 Elizabeth II replaced the Tudor Crown on the lintel with a representation of the actual crown used in coronation ceremonies and, in 1955, Scotland replaced St. Edward's crown with the Crown of Scotland. These iconic structures appear unexpectedly in places far afield from the United Kingdom—Hong Kong, Antigua, and the Caribbean—emblems of an empire on which the sun eventually set. However, until the early eighties, when the British telephone industry was privatized and the Post Office telephones came under the jurisdiction of British Telecom (BT), the basic structure remained essentially the same. "They were not romantic," a British friend of mine emphasizes. "The doors on them were heavy, they rarely closed properly, they were cold and draughty and they stank of piss."[13] Of course, he added, the new ones weren't any better. BT soon launched a more utilitarian design series, announcing that it would be replacing the classic British booth because they "don't make enough to cover their cleaning costs" according to Les King, a public relations leader for BT's pay phone division.[14] In addition to removing booths, BT increased the minimum charge for a local call, and, in 2002, halted its manufacture of phone cards.

In Britain, however, unlike the United States, groups of people successfully campaigned to have the booths protected by the same legislation that protects buildings with demonstrated architectural and historical significance. In 1988, one thousand booths were designated national landmarks. The efficacy of these efforts is apparent in, of all places, the first episode of the BBC's modernized *Sherlock*, in which James Watson is dogged by a ringing pay phone throughout the streets of London. When he finally picks it up, the phone call turns out to be from Sherlock's brother, who, as Sherlock says, "*is* the British government." Nevertheless, in

FIGURE 4 British telephone box, Creative Commons (standard creative commons, licensed for commercial use, no attribution required images of British booths).

Britain as in the United States, phone booths and landlines in general are a dying breed.

The spirit world

"I am telling you this because conversation is a journey," writes Anne Carson, "and what gives it value is fear."[15] The ornateness of the early booths highlighted the significance of telephone calls themselves, a significance that was lost soon after the initial novelty and anxiety about the telephone dissipated. Before the telephone, people living at a distance from each other communicated mainly by letter and in rare instances by telegram. Handwritten letters bore the physical traces of their authors—perspiration, tears, not to mention the author's characteristic hand—and thus became stand-ins for the person's presence. And yet, letters could not be written in the dark in which telephone conversations could take place. What the telephone and its contemporary technologies did to haunting effect was to create a fissure between what was seen and what was heard, a breach that led Freud to find in telephony a useful metaphor for his psychoanalytical practice: the unseen doctor acting as the telephone receiver for his patients' revelations. Marcel Proust, whose *In Search of Lost Time* manages to capture both the novelty and the banality of the telephone, articulates how unnerving this disembodiment could be near the beginning of *Guermantes Way*, when his narrator describes the first telephone call he receives, from his grandmother:

> Suddenly I heard that voice which I mistakenly thought I knew so well; for always until then, every time that my grandmother had talked to me, I had been accustomed to follow what she said on the open score of her face, in which the eyes figured so largely; but her voice itself I was hearing this afternoon for the first time. And because that voice appeared to me to have altered in it proportions from the moment that it was whole, and reached me thus alone and without the accompaniment of her face and features, I discovered for the first time how sweet that voice was. . . . It was sweet, but also how sad it was, first of all on account of its very sweetness, a sweetness drained almost—more than any but a few human voices can have ever been—of every element of hardness, of resistance to others, of selfishness.[16]

Detached from the body, the voice travels and arrives with nothing but itself, disarming the narrator of what he thought he knew the person to be. Although many suggest that we convey more in our faces than in our words, removed from the "open score" of the face, the voice offers the narrator a strange but potentially more intimate representation of his grandmother. The voice is perhaps the least malleable of our attributes, and thus suggests a portrait closer to the truth. Like scent, it seems closer to her essence, which he had never before been able to feel. It becomes a carrier of meaning in its own right, the sound of sense. For Proust's narrator, the singularity of the voice suggests not just the essence of his grandmother, but is an adumbration of her death:

> Was it, however, solely the voice that, because it was alone, gave me this new impression which tore my heart? Not at all; it was rather that this isolation of the voice was like a symbol, an evocation, a direct consequence of another isolation, that of my grandmother, for the first time separated from me. . . . "Granny!" I cried to her, "Granny!" and I longed to kiss her, but I had beside me only the voice, a phantom as impalpable as the one that would perhaps come back to visit me when my grandmother was dead.[17]

In the disembodied voice Proust hears the dead voice, how the voice will come to him, conjured up from memory's technology, when the body that carried this voice has long since ceased to be. Proust's image embodies one of the central paradoxes of the phone: like the radio and phonograph, the telephone could convey sounds without any traces of decay, but ultimately they made the physical absence of people more apparent.

If for some the disruption between what was seen and what was heard articulated and represented the dead's persistent existence in the minds of the living, for others the telephone suggested access to wherever the dead had gone. Marcel's first phone call occurs on a pay phone in a pharmacy.

He was not alone—many people's first phone calls occurred on public phones. Given how fraught talking on the phone initially was, it's not surprising that the structures enclosing such conversations were sturdy and even luxurious. These booths were framing and containing private encounters of the most acute kind, either between intimate human beings, or, tacitly, between human beings and something beyond articulation.

Lighthouse on the highway

Almost as soon as they were constructed, however, phone booths began to be deconstructed. In 1954, as telephone subscriptions increased at an exponential rate, AT&T introduced the aluminum-and-glass telephone booth, offering a structure that was more transparent and durable, although not as useful as a changing closet for Clark Kent or a hiding place for Charlie Chaplin in *The Idle Class*.[18] These "picture window" phone booths were advertised as sleeker and more sophisticated than the original wooden cabinets, affording views of "the mountain ranges of the Northwest" and "the green of San Diego's famed Balboa Park." Many booths were designed to coordinate with their context—a bamboo booth for a Bali Hai restaurant in San Diego, for example, or a pagoda-shaped booth at the entrance to Los Angeles' Chinatown, indicating that their structure in a sense superseded their function.[19] Curiously, the diminished size

FIGURE 5 Glass Phone Booth in Idyllwild, CA. Photo © David Eng, used with permission.

seemed to encourage college students to try to fit as many people as possible into a space pointedly designed for one, a fad prosaically named "phone booth stuffing."

Subsequent to the glass and aluminum booth, in the 1960s AT&T began to develop a series of structures that were only partially enclosed—the drive-up, walk-up, and boat-up designs, as well as "coin shelves"—that were the precursor to the kiosks that enclose most pay phones today. "Telephone users will be happy to have them," writes Cullen Bryant Colton of a new batch of phone booths being designed in Queens, "for the telephone booth, like the drugstore quick luncheon, is very much part of the American desire to do things quickly."[20] People would move blissfully along the highways, unimpeded by traffic, stopping at pay phones and continuing on as gracefully as a relay runner hands off a baton—that was the idea anyway. In fact, when cellular technology was introduced in America, it was immediately popular in Los Angeles, where people spend an inordinate amount of time commuting and for whom the enclosure of the car provides a natural booth.

Few relics of these more futuristic designs remain other than the stand-alone kiosk, but the ad copy for public phones at the time suggests a paradise of convenience and clear communication. "Like a lighthouse on the highway: A thoughtful husband, hurrying home, phones to reassure his wife." "A sputtering car coasts to a stop and two grateful women phone for road service." "The most familiar public phone is already on duty everywhere—giving service and protection 24 hours a day."[21] This is a piece of technology that is not only responsive to our needs but is a portal to our better selves. In this world no wives are anxious, no flat tire left unfixed, no colleagues offended.

There were other reasons besides speed to explain the way the phone booth evolved: glass was cheaper and made the interior of the booth more visible to the public, which ostensibly made it safer. In more scenic areas, glass also afforded the telephone caller a view and the kiosk, lacking any kind of physical enclosure, made the telephone more accessible to the disabled and to people "on the move." But the evolution of the phone booth—first from opaque to transparent, then from protective enclosure to stand-alone kiosk, also suggests something about the evolution of the individual in public, about the exposure we seek and the exposure we dread. When phone booths became transparent they became what *Anchorman*'s Ron Burgundy so aptly described as "glass case[s] of emotion," where what couldn't be heard renders the person on the phone more luminous and more melancholy, an object of our gaze rather than simply the subject of his or her own life.

Who can forget, once seen, the face of Jeanne Moreau in Louis Malle's 1957 *Elevator to the Gallows*, as she stands in a glass phone booth whispering "Je t'aime" to the lover who will

FIGURE 6 *Elevator to the Gallows*, dir. Louis Malle, 1958.

FIGURE 7 *Glengarry Glen Ross*, dir. James Foley, 1992.

eventually kill her husband. Or, so many years later, the sadness of Shelly Levene standing in a New York phone booth on a rainy night, trying to persuade a hapless husband to buy a piece of worthless real estate in *Glengarry Glen Ross*. Or even John Cusack in *Say Anything*, talking on a pay phone to his sister while intentionally standing outside in the rain, because, after breaking up with his girlfriend, the rainy world outside the booth is much more resonant with how he is actually feeling.

Phone booths were one of our first real attempts to create private areas in public spaces; although the literal structures have all but disappeared, we have recreated them in the fixedness with which we stare at our devices. Has something been lost, stolen, sacrificed in this evolution and replacement? If communication could be imagined as some sort of sum total, which it can't, are we communicating more, better, less, worse than we have in the past? Are we getting closer to the heart of the matter or missing it entirely?

6 FORTRESS OF SOLITUDE

"It is my heart-warm and world-embracing Christmas hope and aspiration," wrote Mark Twain in 1890, "that all of us, the high, the low, the rich, the poor, the admired, the despised, the loved, the hated, the civilized, the savage (every man and brother of us all throughout the whole earth), may eventually be gathered together in a heaven of everlasting rest and peace and bliss, except the inventor of the telephone."[1] Although Twain conceded the telephone's utility, and was in fact one of the first Americans to own one, he was wary of its power to carry the human voice over even greater distances than it was already heard. He was not the only one. America had taken more readily to the automobile, which, with the train, was responsible for transporting people farther from each other and necessitating inventions like the telegraph and telephone that would "annihilate" time and space. When the president of Bell Telephone offered Western Union all of Bell's patents for one hundred thousand dollars, Western Union declined without a second thought, finding no use for them.[2] On

August 19, 1905, during the National Conference of the Old German Brethren, held in Flora, Indiana, the Advisory Board ruled that no member of the Brethren could retain a personal telephone in his house, making them reliant on public phones. In explaining the reasoning behind their stark decree, the board stated: "The telephone does have its innocent uses, its admirable qualities, but it also possesses peculiarities that render its convenient presence an obstacle to the attainment and maintenance of several virtues."[3]

Admittedly, the German Brethren were one of the most conservative religious sects in America, but they were not alone in their concerns. As the essayist Eula Biss has noted, even more than the telephone itself, people questioned the concept that everyone in the country could be connected through a labyrinth of telephone lines. Some argued that the telephone poles used to create this complex labyrinth in the sky not only encroached on private property, but that they also destroyed the landscape, obstructing clear and unfettered views. Bell was indefatigable, however, and by 1880 every major city was wired.[4] In the mid-nineteen seventies, some of these lines would be leased for the use of ARPANET, an early iteration of the Internet that connected federal and university computers.

Since its invention, however, telephone companies have worked diligently to make the telephone an integral and necessary part of human life. Telephone advertisement titles such as "Man for the Moment," "Multiplication of Power," "At a Psychological Moment," and "The Voice of

Success" indicate how telephone companies sought to make the telephone inextricable from notions of identity, power, and success.[5] Telephones became conspirators in our lives, participants in our loves and losses, necessarily implicated in the communications they conveyed. Public phones played into this, allowing people to be immensely more reachable than before, able to wield power from a distance, or at the very last minute notify someone about a later arrival. For many it was more affordable to make a call from a public phone than from a private line, and, before private lines became ubiquitous, it was women who inhabited phone booths most frequently, errands becoming opportunities to connect with people outside their homes. Public phones also made people much more accountable, and in the initial years following their invention, questions about "telephone etiquette" provoked continuous anxiety. Being telephonically connected soon became a personal responsibility and a moral imperative, a condition we are intimately familiar with now but which in the past was so potentially burdensome that in 1962, when most households possessed phones, James Thurber could write that many parents successfully avoided salesman and bill collectors by having their children answer the phone while they continued to enjoy a martini: "Most American parents will not answer the telephone when it rings, but will let a little boy do it."[6]

By the end of the Second World War, half of American houses possessed telephones; consequently pay phones in pharmacies, banks, stores, and street corners remained nexus points for communities, subtly appealing to a sense of civil society in which we are never out of the range of each other's voices. When soldiers started returning from the war, banks of phone booths were erected in preparation for their arrival. On the morning of October 17, ten of the most legendary warships to serve during the Second World War materialized through the morning mist and docked in New York Harbor. The USS *Monterey* appeared first, followed by the USS *Enterprise*, a carrier that had bookended the American war effort against Japan, sending planes to fight at both Pearl Harbor and Okinawa, collectively sailing more than 275,000 miles in a little less than four years. Neither sailors nor spectators cheered as the *Enterprise* landed, "coming home," as Robert Richards wrote, "with a record too obvious for flattery" and "too many ghosts aboard to cause reckless cheering."[7] Nevertheless, when the ships berthed, servicemen descended in a swell, immediately swarming around the telephone booths and the telegram desk the Western Union Telegraph Company had organized outside the warehouse. They were, as the *New York Times* reported the next day, "eager to call their wives, mothers and sweethearts from coast to coast."[8]

How many phone calls were made, what was said—these details are lost. What remains is the indelible image of contact. There's a ceremony and romance to communication here that only emphasizes how much the nature of communication has changed. While soldiers serving in the Second World War relied primarily on letters and the odd telephone call made from a public phone as their means of maintaining contact with home, often enduring months without any contact at all, soldiers leaving for Iraq and Afghanistan in recent years were encouraged to bring "semi-rugged" laptops, so that they could take advantage of the numerous Wi-Fi hotspots the US government has spent millions of dollars building. Once connected, soldiers have a dizzying array of options: e-mail, FaceTime, Skype, instant messaging, to name only a few.

"It's rejuvenating," said Sergeant Grelak to the *New York Times*, speaking about the ability to stay closely connected to his family. "It keeps you from getting detached from the person you left behind. You go outside and you run the risk of getting shot and blown up. That changes people. If I didn't have that connection, I would feel like a stranger."[9] Although Sergeant Grelak's situation is more extreme than most of our daily experiences, the relationship he posits between identity and connection is important and partly explains our complicated dependency on our telephones, a reliance that public phones facilitated. Part of what is so moving about the image of soldiers returning home from the Second World War is the acuteness of contact, of the sensation that people *had* changed and become strangers, and that this was the beginning of the reckoning with that strangeness.

Between 1997 and 1999, just as cell phones were becoming more commonplace, there was a surge in pay phone usage. One can only speculate as to why, but one reason might be the tacit pressure to remain connected and in range exerted by cell phones. In the early days of the cell phone, like the early days of the telephone, most of the people who owned these devices were wealthy or in business. It was a mark of status and importance to be seen using a cell phone because the people were involved in business and communication that couldn't wait. By 2014, cell phones had become ubiquitous: 58 percent of the American population possessed smart phones and 90 percent possessed cell phones.[10] As early as 2008, there were more mobile phones in the United Kingdom than there were people. Unlike the world Thurber depicts, people now are so loath to be separated from their devices that they appear like additional appendages, evolutionary add-ons to the ear and the hand. Nevertheless, many speak about their current relationship with technology in language usually reserved for addiction, opining on the various ways technology seems to have infiltrated every aspect of their lives. Significant numbers are beginning to engage in what seems to be religious restriction of their phone usage, referring to phone and e-mail Sabbaths and Shabbats, congratulating themselves on their powers of discipline and abnegation when they can endure twelve hours without getting a dopamine-inducing dose of e-mail, Facebook, or Twitter.

In his 1987 essay The Ecstasy of "Communication." Jean Baudrillard asserted that our relationship to the material world was attenuating, and that objects were being replaced by surfaces and interfaces: One of the myriad consequences of this is that boundaries—between private and public, subject and object—also diminish; rather than individuated selves, we become networks. At the end of his essay, Baudrillard pays particular attention to the schizophrenic, in whose disorder Baudrillard finds a representation of the individual who is assaulted by the forces of modern life: "It is the end of interiority and intimacy, the overexposure and transparence of the world which traverses him without obstacle. He can no longer produce the limits of his own being, can no longer play nor stage himself, can no longer produce himself as mirror."[11] As the world traverses us unimpeded, our identity is no longer characterized by wholeness but by fragments.

In this context, what the scholar Avital Ronell refers to as "the carceral silence of a telephone booth" has a curious appeal.[12] As phone companies systematically remove pay phones, Amish and Mennonite communities have been building, or rebuilding, their own. In 2006 the *Washington Post* reported on the erection of twelve phone booths (made of recycled materials like oil tanks) in remote areas of a Mennonite settlement in southern Maryland. Now these religious followers can use the phones to communicate about important business or personal matters, but also adhere to the sanctions that prohibit individual ownership of technology such as radios, televisions, automobiles, and telephones. Referred to as "phone shanties" and hidden in the woods, behind barns and chicken coops, these "community phones" are intended to isolate contact with the external world and lessen the potential for such contact to divert people's attention from faith, family, and community.[13]

FIGURE 8 An Amish phone shanty. Photo © Lori Garske, used with permission.

These revisions of an obsolete technology underscore their symbolic and functional value. Every telephone call is haunted by the uncertainty of what it will ask, what news it will bring. "You don't know who's calling or what you are going to be called upon to do," writes Ronell, "and still, you are lending your ear, giving something up, receiving an order."[14] The telephone possesses the power to command from a distance, and has therefore been an instrumental tool of authority, from totalitarian regimes to parents. From its interior sound results, orders, information, rendering us powerless before a machine that sometimes seems like it is the arbiter of our fates. The phone call in the middle of the night, the unexpected call during the day, continues to plague our dreams, caller ID notwithstanding. Horror films capitalized on this inherent uncertainty, as did prank callers, whom horror films often featured. A "cool medium," according to Marshall McLuhan, because it provides less information and requires more participation, the equivalent of a seminar as opposed to a lecture. A disturbing number of people, it turns out, have committed suicide in phone booths, often after hanging up the telephone. When James Conway, in Martin Scorsese's *Goodfellas*, gets the news in a phone booth that his good friend Tommy has been whacked instead of being made in the mafia, all he can do is destroy the messenger, repeatedly banging the receiver and finally kicking over the booth itself.

FIGURE 9 *Goodfellas*, dir. Martin Scorsese, 1990.

FIGURE 10 *Anchorman: The Legend of Ron Burgundy*, dir. Adam McKay, 2004.

While the telephone itself seemed to transcend some of the bounds imposed on us by nature, the telephone booth acknowledged and conceded to them. Even the new Facebook campus in Menlo Park, California, has numerous phone booths scattered across the fifty-nine-acre campus where people can make their phone calls. The phone booths are empty, of course, but they promise silence and privacy in a community where not even the chief executive has his own office and whose central product by all accounts is one of the most psychologically intrusive and noisy products of the twenty-first century. The booth neutralizes the mobility offered by the modern phone, enforcing stillness. The Romanian religious philosopher Mircea Eliade has suggested that structures erected in defense of cities—moats, labyrinths, and ramparts—most likely originated as defenses against "demons and the souls of the dead" rather than armies. He describes, in northern India, how a circle would be inscribed around a village during an epidemic in order to prevent sickness from entering, and how, in Europe during the Middle Ages, the walls that surrounded each city would be "ritually consecrated against the devil, sickness, and death."[15] Each demarcation attempts to preserve order against chaos. In much the same way, the phone booth suggested the need for safeguards against threats less articulable than noise and weather. We have been taught that the subject under observation changes due to this observation, and that even the most tangential contact has the power to linger within and change us. Modern life is in fact premised on the intensity and rapidity of contact. "The psychological foundation upon which the metropolitan individuality is erected," writes Georg Simmel in his essay "The Metropolis and Mental Life," "is the intensification of emotional life due to the swift and continuous shift of external and internal stimuli."[16] In such a world the phone booth was a welcome reprieve, a space in which one could gain breathing room and that was a poignant acknowledgment of what progress has cost us.

In *The Misfits*, from 1961, we first meet Montgomery Clift's Perce when he is waiting for a call from his mother outside a phone booth in the middle of the Nevada desert. It's a curiously affecting scene—Marilyn Monroe's Rosyln Tabor and Clark Gable's Gay Langland are watching and listening from the car. We learn that Perce has just been in

Colorado, where he won a hundred dollars, which he did not send to his mother, apparently, but instead used to buy a pair of boots, and that he's recently been injured, but that he has since healed and also, at the very end, that his mother has remarried someone he doesn't like. The operator ends the call before Perce has time to reply adequately, at which point he hangs up and catches a ride with Rosyln and Gay to the rodeo. It's a tender moment, one that presages a scene of true intimacy, in which, after sustaining another head injury, Perce tells Rosyln about how the ranch his mother had promised him had instead gone to his stepfather, making explicit the loneliness that so clearly infuses his actions.

The soldiers returning home from serving in the Second World War might have been rendered strangers to themselves and to their families by war, but the phalanx of booths intimated what was necessary for their return. In *Local Hero*, a 1983 film in which an oil executive travels to a remote coastal town in Scotland in order to negotiate a deal for the sale of oil-rich land, there is one working telephone booth that is situated on a hill with a breathtaking view of the ocean. The residents routinely clean the booth, even giving it a fresh coat of paint, highlighting the importance not only of the connection it provides, but also the sanctuary in which to make that connection.

7 SIGNIFICANT PORTALS

The creek was made narrow by little green trees that grew too close together. The creek was like 12,845 telephone booths in a row with high Victorian ceilings and all the doors taken off and all the backs of the booth knocked out.

RICHARD BRAUTIGAN

Even without booths, telephone conversations create locations, circumferences of absorption in which we sit, stand, circle, pace, gesticulate, think, and feel. Like the spaces we inhabit as readers, they are locations in which we are both here, and elsewhere—I can be staring at something or someone, but if I am engrossed in the conversation I will see only an internal landscape, what is visible to me and no one else. Trying to retrieve the features of this landscape after the telephone call is over is a fool's errand—it disappears as a dream disappears when we try to remember it.

But we were once much more tethered to the earth. The evidence was everywhere in the form of cords roiling in corners like snakes, or splayed across the floor like tripwire—from the receiver to the phone, from the phone to the wall, from the wall to the poles, from the poles to the switchboard. In the late 1870s and early 1880s, before Bell learned how to consolidate connections, there were so many phone lines weaving through the sky one could look up and see only slivers of blue. These lines were connected to a central switchboard run by thousands of mostly female operators, before the system became largely automated in the 1940s. Numbers came to include area codes, and area codes became, for some at least, a certain kind of identity marker. People in New York City and its boroughs, for example, were known to put up a good fight in order to retain Manhattan's "212" area code, even when they lived in an outlying borough like the Bronx or Queens. These were the days, which lasted well into the nineties and early aughts, when each of us knew at least two or three numbers by heart. Now many of us have area codes not from where we live now, but from some place we lived earlier—an earlier stage of life preserved in numerical amber.

Before many of us acquired the mobile phones we carry with us at all times, set beside our beds before we go to sleep, and pick up again when we wake up, our telephones occupied specific, intentional places where we lived. Perhaps the telephone rested behind a staircase, at the end of a hallway, or more centrally, on a desk in a living room or on a wall in the kitchen. Perhaps it lived at the juncture of the kitchen and the living room, or, as one friend described, while the telephone hung officially in the kitchen "with an epically long phone line," it could stretch downstairs to the den, or around the corner to the dining room. In houses with multiple phones there was usually a hierarchy, the most important residing in the most central but least private place, leaving room for the "telephone paraphernalia" of notepads, pencils, and pens.[1]

Where a telephone was placed told us something about the person who had placed it, and maybe something about the lives they were living. One friend poignantly remarked that after her parents divorced there were so many moves and houses, the locations of individual telephones no longer stood out in her memory. Another wrote that in her house there were more and more phones as her brothers started getting into more and more trouble, requiring daily telephone calls to appease one harassed adult or another. Still another remembered that the family phone lived somewhere between the living room and the kitchen, "augmented with hella long ass handset cables—upwards of 10-15' which could be extended to a desperate 20' so you could watch MTV. . . . Cordless was for fucking yuppie assholes who didn't understand the relative value of stringing a 50' cable."[2]

The telephone of my own childhood lived above the couch, kitty-corner from the bookcase, not a private space but its own space, and one that never migrated as Walter Benjamin describes the location of the phone in his childhood house migrating over the course of time to a more privileged place of attention: "The apparatus, like a legendary hero once exposed to die on a mountain gorge, left the dark hallway in the back of the house to make its regal entry into the cleaner and brighter rooms that were now inhabited by the younger generation."[3] Even when it inhabited the servant's quarters of the hallway, however, Benjamin describes how the telephone "augmented the terms of that Berlin apartment with the endless passage leading from the half-lit dining rooms to the back bedrooms."

The telephone's ability to "augment the terms" of any space it inhabits is perhaps why the Japanese, in the early aughts, were working on technology to make enable phone booths to transmit holographic images. It is also perhaps why phone booths have been variously appropriated by cinema as portals to other dimensions and time-travel machines to different eras. On March 31, 1999, the Wachowskis released *The Matrix*, a film that depicts a dystopian world in which humans have irrevocably damaged the earth and "reality" is actually a simulation—the Matrix—created by code and energized by humans' body heat and electrical activity. The hero of the Matrix is Neo, a young computer programmer who works in a nondescript cubicle in an anonymous city. Spending his off hours hacking into various computer systems, Neo is eventually recruited by Morpheus, the leader of a rebel force who has managed to escape the Matrix and who presents Neo with the alluring possibility of maintaining the status quo: "You take the blue pill, the story ends, you wake up in your bed and believe whatever you want to believe," or experiencing truth: "You take the red pill, you stay in wonderland, and I show you how deep the rabbit hole goes."[4] Neo chooses to go down the rabbit hole and realizes that he (and everyone else) are inert batteries, fueling a vast computer system. Morpheus and his rebel force live in an underground cell where the food is dismal and people are dressed in rags, a disappointing reality obviated by what Morpheus and his crew have learned how to do: master the Matrix and achieve autonomy in a world that is controlled by a computer program. Morpheus contacts Neo because he believes Neo is the one who will not only be able to see the Matrix for what it is, but also free everyone else from its clutches.

Although cell phones of a kind are present in *The Matrix*—both the rebels and the sentient machines use them to contact each other—the rebels can only enter and exit the Matrix through landlines, most often in phone booths. When asked why this was the case, the Wachowskis replied, "Mostly we felt that the amount of information that was being sent into the Matrix required a significant portal. Those portals, we felt, were better described with the hard lines rather than cell lines. We also felt that the rebels tried to be invisible when they hacked, that's why all the entrances and exits were sort of through decrepit and low traffic areas of the Matrix."[5] At the end of the film Neo announces his intentions to the Matrix from a phone booth on a busy street: "I'm going to hang up this phone, and then I'm going to show these people what you don't want them to see. I'm

FIGURE 11 *The Matrix*, dir. Andy and Lana Wachowski, 1999.

going to show them a world without you, a world without rules and controls, without borders or boundaries, a world where anything is possible. Where we go from there, is a choice I leave to you."[6] Although the sequel to *The Matrix, The Matrix Reloaded*, was released just four years later in 2003, phone booths were gone, already defunct as a means of communication and transportation.

Portals are elliptical passageways, eliding the journey between one place and another. In *The Matrix* this elision resonates with how Morpheus emphasizes that "reality" is really a matter of perception, easily manipulated if we are only strong and skillful enough to do it. Portals permeate everyday life now, but they are mostly in the form of websites, arguably the most mysterious and simultaneously most banal passageway we have. In contrast, landlines and phone booths were objects commensurate with the leaps people were making, akin to wardrobes and rabbit holes. As mysterious they were, these portals had heft, substance, materiality. For many of us, telephones were our initial portals to private lives, and, for those of us of a certain age, telephones were our first portals to the Internet, when we had to dial up the ether rather than be connected instantaneously and wirelessly. In 1995, when the Internet first came to widespread public consciousness via Netscape, Mark Thomas started a website called The Payphone Project where he listed hundreds of numbers from pay phones and encouraged people to call random numbers, creating chance encounters. Coincidences proliferate; meaning accrues. But when pay phones increasingly started to allow only outbound calls Thomas turned the website into a pay phone museum of sorts, gathering images, articles, and stories that pertain to the public phone and thereby providing an invaluable archive.[7]

Undoubtedly the most famous telephone booth is that used in *Doctor Who*, a British TV series in which a Doctor of Time travels through different eras in a repurposed British police box called a TARDIS. An acronym for Time and Relative Dimension in Space, the TARDIS is a piece of technology created by the Time Lords. One of the many capabilities of the TARDIS is the ability to transform to suit its circumstances. However, the TARDIS has a faulty chameleon circuit, and consequently has permanently taken the form of a 1960s British police box. Although it was probably cheaper to keep the Doctor's time-travel machine a police box rather than spending the money to develop a new one for every show, which ran weekly, the choice of a police box in the first place is significant. Jill Lepore finds a possible reason for it in the history of beat policing, which arose in nineteenth-century England, when police officers controlled specific territories and developed strong relationships with the communities for which they were responsible. "Doctor Who polices worlds. The idea of a world's policeman dates to the First World War and began to come into common usage near the end of the Second," writes Lepore. "In 1943, during a birthday dinner for Winston Churchill, F.D.R. called upon the allied powers—the United States, Great Britain, the Soviet Union, and China—to serve as the world's 'four policemen.'"[8] *Doctor Who* premiered in 1963, at the tail end of the British Empire and in the heart of the Cold War, a time when the efficacy of Britain as a "world policeman" had faded. The Doctor of Time is the closest thing Britain has to a superhero,

FIGURE 12 *Doctor Who*, dir. Joe Ahearne, 2005.

but, unlike most superheroes, he feels the weight not only of personal history but the history of the entire universe he traverses so easily with the help of his telephone box.

The decision for Alex Winter's Bill Preston and Keanu Reeves's Ted Logan in *Bill and Ted's Excellent Adventure* to travel through time from their home in San Dimas, California, in a phone booth was more circumstantial than symbolic—the director had originally planned to use a car but was worried about appearing as if he was appropriating *Back to the Future*—but the booth remains one of the movie's most memorable and endearing features. Bill and Ted are in danger of flunking their history course unless they can deliver "something very special" in their oral report due the next day. Unbeknownst to them, the fate of the world depends on their successfully passing history and thus assuring the sequence of events that leads to the utopian society that exists in 2688. Consequently, Rufas, an emissary from the future, arrives in

FIGURE 13 *Bill and Ted's Excellent Adventure*, dir. Stephen Herek, 1989.

a phone booth to accompany Bill and Ted through time as they gather "research" for their presentation.

To a certain extent this repurposing of the phone booth offers one of the most accurate representations we have of what the phone booth actually means to us. Utterly quotidian, the phone booth nevertheless had the ability to transport us out of the world in which we were living. That this representation is offered in a film that has acquired a cult following among stoners and other refugees from the nineties tells us how deeply this understanding of the phone and phone booth really goes. What's more, to see a glass and aluminum phone booth appear in Napoleonic France, the old West, or in ancient Greece is not much more anachronistic than encountering a booth on a street in contemporary life.

Rootlessness

Obsolescence has remade phone booths into real time-travel machines. Most of those that remain suffer the same fate as the Ramsays' estate in Virginia Woolf's *To the Lighthouse*. Abandoned by the family, the house is overtaken by nature: "Only the Lighthouse beam entered the rooms for a moment, sent its sudden stare over the bed and wall in the darkness of winter, looked with equanimity at the thistle and the swallow, the rat and the straw. Nothing now withstood them; nothing said no to them."[9] The functional phone booths that interrupt stretches of the Pacific Coast Highway at regular intervals are perfect examples of how nature is reclaiming technology receding like deadfall back into the world from it sprung.

A few miles south of Bodega Bay, where Tippi Hedren's Melanie Daniels famously sought shelter in a phone booth from an ever-increasing flock of venomous birds, is a phone booth that sits beside a restaurant looking out to the Pacific. Rust has corroded parts of the back panel that was once a deep sky blue. On one of the outside corners, a handmade sticker for Doomsayer is peeling back from its edges and various hieroglyphs have been scrawled across amenable surfaces. Sea grass and weeds are growing up through the cracks in the cement base, finding their way in the gap between the walls of the booth and the ground. In the distance, a lone schooner floats against the horizon, perfectly framed by the glass panes of the booth. The manager has been embroiled in a persistent battle with the telephone company that wants to remove the phone because it has long since stopped making anything close to a profit. "It's true, it costs more money to maintain than it makes," he tells me when I ask him about the booth, "but this whole stretch of the PCH is a dead zone. That phone booth is a civic service."

When people discuss the consequences of the disappearing pay phone, though, the two groups that come up most often are immigrants and the homeless, categories of people defined by their relationship with place, or lack thereof. "Pay phones are lifelines for the down and out," Mark Thomas commented to the

New York Times, "their booths are rainy-day cocoons," he said. "You lose those, and you lose a lot of windows onto the human condition."[10] Although the number of working pay phones available in the United States continues to rapidly diminish, many remain in areas with high immigrant populations such as Columbia Heights in Washington, D.C., or East Los Angeles. The advertisements that cover the surfaces of these pay phones tend to be for prepaid international calling cards as much as for taxi services and call girls. According to Robin Harris, the president of Robin Technologies, a corporation that owns and operates pay phones in the Washington, D.C., area, "Immigrants are one of the niches that the pay phone community serves very well. If someone wants to call home, say to El Salvador, it can actually be less expensive for them to use a public phone than a cell phone."[11]

Most homeless people do not have access to any telephones but public ones, and yet few provisions have been made for them in the event of pay phones being removed entirely. In 2011, Contexture Design, a design firm based in Vancouver, B.C., released a repurposed phone booth that contained shelves, a radio, a sink, and a seating area that unfolds into a bed. Contexture's website explains the project as an "exploration of shelter and the concept of home, particularly as it relates to homelessness," while it also emphasizes that their repurposed booth is not a "realistic solution." Nevertheless it does acknowledge the fact that for many homeless, the phone booth is far from dysfunctional.

FIGURE 14 Contexture shelter, taken from the Contexture website.

Contexture's redesigned phone booth is a practical response to its increasing obsolescence, but most repurposed phone booths do not strive for practicality, perhaps because phone booths were marketed as tools of convenience and emergency, but always as tools, emphasizing what Martin Heidegger might have deemed their "readiness-to-hand." When we use a phone booth we don't interrogate its identity; we simply use it, more or less automatically. It's only when an object breaks, or is otherwise rendered useless, that it becomes characterized by something other than its utility. The graffiti artist Banksy made this point quite well when he installed a broken phone booth on a street in Soho. Knocked on its side, broken nearly in half, a pickax sticking out from

FIGURE 15 Banksy's Broken Booth. Photo © Nick Cunard, used with permission.

one of its walls and blood spilling from beneath it, Banksy illustrated the willful murder of the phone booth.

We are now much more likely to encounter a phone booth in which we can call someone to hear a poem, or the recording of a beautiful violin concerto, than we are a phone booth in which we could call our spouses. In 2008, the one phone booth and mobile lending library disappeared at the same time from Westbury, a small village in the county of Somerset, England. Instead of letting it go, the residents collectively bought the phone booth and turned it into a lending library. (The project was repeated in 2012, when the architect John Locke began to turn abandoned phone booths throughout New York City into lending libraries.) In

FIGURE 16 Phone Booth Aquarium. Installation by artist
Benedetto Bufalino, in conjunction with Benoit Deseille. Photo ©
Nicolas Nova, used with permission.

2011, Kingyobu (the Goldfinch Club) repurposed a group of
abandoned phone booths as goldfish aquariums and started
installing them around the city of Osaka. A phone booth
in Yellow Springs, Ohio, was the site of twelve different art
installations between 2009 and 2010. The Wi-Fi hotspots
all but three remaining pay phones in New York City will
become—which look like enlarged smart phones—are called
"links." So the portals remain, but they are taking us to
different places, or perhaps they are the same places strangely
configured.

8 A FINE AND PRIVATE PLACE

"People have a right to their secrets," says Valentine in Krystof Kieslowski's *Red*.[1] She is talking to a retired judge, who spends his leisure time listening to other people's telephone conversations, imploring him to stop. Central to most Americans' accounts of where the telephone lived in their houses is the tension between the need for accessibility and the desire for privacy. Households can be refuges and retreats from the world, but they can also be the last places to find any sort of privacy. The only phone in my aunt's house was "as central as possible: in the dining room, which was in the middle of the house, between the kitchen and the living room," while my uncle, who grew up above the grocery store his parents ran, had no phone in his house and was forced to use the pay phone in the grocery store until 1959, when supermarkets took away too much business for their local store to remain solvent.[2] The phone was enclosed in a wooden booth and thus afforded more privacy than most, but he doesn't remember even wanting privacy until love became factor in his life. Suddenly it was no longer conceivable for telephone calls to be public events, vulnerable to scrutiny. "I am waiting for no more than a telephone call," writes Roland Barthes of waiting for a lover to phone, "but the anxiety is the same. Everything is solemn; I have no sense of *proportions*."[3]

Privacy is the ability to remove one's self from the fray. It affords breathing room, a space in which to reflect. A Korean woman mentions to me casually that when she was growing up, in a very strict household, she went down the street to the pay phone at the local gas station to talk with boys—it was the only place she could get any privacy. Another friend describes moving to Seattle from New York and having therapy sessions with her New York therapist over the phone. "My husband and I were splitting up but still living together. I was really miserable at the time, and I'd go to Elliot Bay. Sometimes three times a week, and in their basement were these great old wooden telephone booths—you could sit in them, and close the door (a hinged double door) and it was sound proof."[4]

When Gray invented the pay phone, it was already instinctual to place phones in quiet places, in corners of rooms, or even in their own enclosures, sometimes referred to as "silence cabinets." In my great grandmother's house on Cape Cod, built around 1895, there was a small telephone room off the entrance hall and adjoining the coat closet. It was booth-sized and included a seat, and by the 1950s it remained the only telephone in the house. In an effort to prove the telephone's practicality, a demonstration at the Paris Electrical Exhibition in 1881 transmitted the sounds of the Grand Opera through the speakers of eighty telephones situated in what were called telephone rooms, large rooms lined with telephones and chairs, where each individual could pick up a telephone receiver and listen to the opera without being disturbed. Conducted every night between 8:00 p.m. and 11:00 p.m., the exhibition was enormously popular. "Certainly nothing has ever been done before so effectually to popularize science," reads an 1881 article in *Scientific American*, "and to render the masses familiar with the effect, however ignorant they may be of the cause, of this marvelous invention, the first feeble voice of which was heard in the Centennial Exhibition of 1876."[5] Of course, the conversation conveyed through these lines was one-way, but the intention to create multiple private worlds able to coexist without intruding upon each other was clear.

We seek privacy intuitively with our mobile phones when we retreat to secluded, isolated, even beautiful places for our conversations. Bob Jacobs, an architect and husband of the famous urban theorist Jane Jacobs, was acutely aware of the need for privacy and installed a centrally located phone booth in his house. Others who also had the means or space to do the same made makeshift booths, creating privacy by any means possible. "In our house in Reno," writes one friend, "the telephone lived in a small antechamber/hallway that connected two bedrooms and the bathroom. . . . It was an in-between room, but you could close all the doors leading to or from it (5 doors total!) for some semblance of privacy—a semi-phone-booth effect."[6] Another friend declared that the most important phone resided in the kitchen, in plain view, but that he could make an ersatz booth for himself by running "the cord through the doorjamb and standing with the door to the kitchen closed, on the top step of the flight to the basement."[7]

In 1890, a year after Gray patented the phone booth, two lawyers from Boston, Samuel Warren and Louis Brandeis, published the most famous formulation of privacy in the *Harvard Law Review*. Titled "The Right to Privacy," Warren and Brandeis argued that people had a right to "inviolate personality." To violate an individual's privacy, they argued, was commensurate with violating their souls. Integral to their argument was the fact that this right was not innate, but instead was a consequence of technology that gave people unprecedented access into each other's lives. "The intensity and complexity of life, attendant upon advancing civilization," Warren and Brandeis wrote, "have rendered necessary some retreat from the world, and man, under the refining influence of culture, has become more sensitive to publicity, so that solitude and privacy have become more essential to the individual; but modern enterprise and invention have, through invasions upon his privacy, subjected him to mental pain and distress, far greater than could be inflicted by mere bodily injury."[8] Warren and Brandeis were referring specifically to cameras and newspapers, but their critique is perhaps even more resonant today, when we are surrounded by technology with which we can expose ourselves, let alone be exposed by others.

As the middle class developed in nineteenth-century America, so too did an interest in privacy and its physical manifestations, as well as in the ways that privacy could potentially be violated. When phone booths were first patented, few individual households possessed phones and those that did usually had party lines. Unsurprisingly, the documentation of early telephone use was peppered with anecdotes of various kinds of betrayal, an inevitable consequence of being privy to each other's intimate conversations. Woody Allen's *Radio Days*, which recounts the early years of the radio, is full of people listening to each other's conversations through party lines. Consequently, phone booths carved out a private space in the public sphere, allowing us to do for the first time what most of us unwittingly now do every day when we speak absorbedly into our personal devices or simply into the air—behavior that we once found easy to describe as insane. Still, people were advised to be circumspect, even in telephone booths. "Never talk private affairs over a public telephone, unless you are sure that the booth is sound proof. The girl who quarrels with her fiancé in the corner drug store and the wife who berates her husband for not coming home on time to dinner, at the grocery store where she trades, belong in the same ill-bred class."[9]

Booths were not the only solution to the need for privacy. In the 1920s, a man named Henry Tuttle developed a device called the Hush-A-Phone, a bauxite cup that could be placed over the speaking end of a receiver, preventing anyone else from hearing what the speaker was saying. The company promised that, with the Hush-A-Phone, each phone could be its own booth, and Tuttle did reasonably good business until the 1950s. In the 1940s, though, AT&T launched a lawsuit against Tuttle, arguing that the Hush-A-Phone lessened the sound quality of telephone conversations and that, moreover, Tuttle's invention violated the agreement AT&T had reached with the government that essentially stated any telephone-related device designed by someone other than AT&T was illegal. The heart of Tuttle case was that the Hush-A-Phone offered something vital that AT&T did not. Mental health and privacy were what the Hush-A-Phone guaranteed, Tuttle argued, and what nothing else offered comparably well. The trial was held in 1950, but it took the FCC (Federal Communications Commission) five years to make a decision, ultimately ruling in AT&T's favor, a decision which Tuttle successfully appealed, but by that time no one was interested in the Hush-A-Phone.[10]

Although it never became a household item, the Hush-A-Phone responded to one of the central tensions of a phone booth. The booth, which has all the appearances of being a place of secrecy—like wardrobes, drawers, and chests—is in fact a public space that offers twenty-four-hour access to anyone. Unlike other public spaces, though, public phones are not devoid of characteristics. Users leave traces of their identity that are not washed away but simply layered over by other traces. Gradually, the personal traces accumulate to such a degree that they effectively cancel each other out, much in the way the dead are defined less by individual lives lived than the fact that these lives are over.

"Insidious tentacle"

When Howard Hughes became a regular at the Beverly Hills Hotel, in 1942, he was in his late thirties and already one of the wealthiest people in the world, famous as a film producer, aviator, and entrepreneur. He was also becoming infamous for the obsessive-compulsive disorder that would intensify and come to dominate his life over the next several decades. The disorder manifested itself in myriad ways, most readily in his distaste for being around other people, exacerbating an already eccentric personality. Toward the end of 1947, after Hughes survived the first of several near-fatal plane crashes, he locked himself in the screening room of a studio near his home for four months, reputedly living off of chocolate and chicken, distracting himself from near-constant physical pain by watching movies. Soon after Hughes emerged, he rented a bungalow in the Beverly Hills Hotel, reserving individual rooms for several associates and girlfriends. He put the hotel staff through their paces, demanding that they place roast beef sandwiches in the crook of a particular tree, hide pineapple upside-down cakes for him throughout the grounds, and, finally, install a phone booth in his suite. The hotel, of course, had personal phones as well as public phone booths, but by that time Hughes felt that most of the world could not be trusted. "They'd switch different booths in and out of different bungalows," reported producer Richard D. Zanuck to the *Los Angeles Times*, "because he [Hughes] didn't want to go through the hotel operator."[11] Hughes had the requisite power and money to remove himself from the fray as much as he wanted, enjoying both the psychological and physical shelter provided by a literal phone booth.

Before he became a guru and "godfather of the New Age," famous for the twelve books he wrote based on his purported interactions with an Indian shaman named Don Juan, Carlos Castaneda attempted to write a nonfiction book about telephones called *Dial Operator*. Castaneda's wife at the time worked as an operator for Bell Telephone, and together they would host parties to discuss philosophy and politics and also to practice ESP. As Mr. Castaneda went on to be deified and vilified by public opinion at differing turns, however, he became increasingly paranoid. Only a small inner circle ever knew of his physical whereabouts; he never gave out his address and only made calls from pay phones.

Revealing representatives of their time, Castaneda and Hughes are often cast as particularly American creations. Brilliant men, at least some of whose success depended on and then ended up perpetuating their mythological biographies, at some point they began to feel more besieged by the world than conquerors of it. For both, the telephone itself was a particularly dangerous technology. "The telephone remains paradigmatic," writes Erik Davis in *Techgnosis*, "since the mere possibility that unknown and unseen agents are bugging your line is enough to puncture the psychological intimacy affected by a phone call, transforming your humble handset into an insidious tentacle of unwanted and invisible powers."[12] The original operators that Hughes so distrusted were gradually replaced by an automated system, in part for efficiency and in part to preserve privacy. While telephones have often been the agents of tyranny, the ability to connect laying just this side of the ability to command, they have as often as not been used to disempower people. Phones could be bugged, tapped, and phreaked, and they often were.

Public phones were harder to trace, harder to overhear, harder to tap, and the booth itself afforded people a degree of privacy that was constitutionally protected. In 1967 a man named Charles Katz was charged with "illegal gambling across state lines" by the federal government. The government had obtained the necessary evidence to charge Katz by placing a warrantless wiretap on a pay phone Katz was known to use to conduct his business. There was some precedent for the legality of this as, in 1928, the Supreme Court had ruled that the use of a wiretap to discover whether someone had violated Prohibition was indeed legal, because the taps had been placed on the phone lines outside the defendant's house rather than within it, where he might reasonably expect privacy and thus did not violate the Fourth Amendment. Writing for the dissenting side, however, was Louis Brandeis, who presciently argued that "the progress of science in furnishing the government with the means of espionage is not likely to stop with wiretapping. Ways someday may be developed by which the government, without removing papers from secret drawers, can reproduce them in court, and by which it will be enabled to expose to a jury the most intimate occurrences of the home."[13] In 1933 Congress responded to this ruling by passing the Federal Communications Act, requiring warrants for taps, and in 1961 the court amended its 1928 ruling by stating that an unlawful search would happen if a "constitutionally protected area" is invaded. Katz argued that a public phone booth was a constitutionally protected area and eventually won the case. "For the Fourth Amendment protects people, not places," stated the ruling. "What a person knowingly exposes to the public, even in his own home or office, is not a subject of Fourth Amendment protection. . . . But what he seeks to preserve as private even in an area accessible to the public, may be constitutionally protected."[14]

Francis Ford Coppola's 1974 film *The Conversation,* however, highlights how paranoia could sometimes be facilitated by use of public phones. Harry Caul, played by Gene Hackman, is a surveillance expert in San Francisco hired to record a couple's conversation. A scene at a conference advertising the increasingly sensitive and sophisticated surveillance technology that is available seems like a harbinger for our contemporary situation, and Caul's awareness of how much is perceivable, in spite of our best efforts to conceal it, makes him hypervigilant about preserving his own anonymity. Like Hughes and Castaneda, he makes virtually all of his calls from public phones. It's difficult to contain this kind of vigilance, however, and it has unintended consequences. Caul's paranoia makes it hard for him to maintain intimate relationships and, what's more, makes him vulnerable to misconstruing what he observes and hears. When he is hired to record a couple's conversation, he starts to become invested not just in getting the recording but in understanding what people are saying, fearing for their safety as well as his own. Caul's fears are misplaced, however; and because he has misunderstood what he's heard, he is unaware of who is truly in danger.

FIGURE 17 *The Conversation*, dir. Francis Ford Coppola, 1974.

Crimes and misdemeanors

There is something ironic about how Hughes and Castaneda sought privacy and anonymity through public phones, only to have this obsession itemized as a revealing detail of their identities. Walter Benjamin wrote about his father assuming a personality on the telephone that he assumed nowhere else. Public phones offer even greater opportunities for personas. You take liberties in phone booths; you are, sometimes, not quite yourself. Phone booths return our anonymity to us, in the way traveling in a foreign country can. It's impossible to imagine film noir, gangster films, and thrillers without phone booths and pay phones. Phone booths are as much a period prop of thrillers and noir as Venetian blinds and hard liquor. It is in the films and novels of these genres that the soldiers returning from the Second World War who called their loved ones from phone booths set up along the shore of the North River found themselves.

The phone allowed liaisons to be arranged, crimes to be planned, help to be sought, and crucial information to be imparted, while the phone booth often acted as a stage within a stage, framing and dramatizing numerous encounters. Without public phones, and before the ubiquitous mobile, most narratives would not have made it past the first plot point. Ray Milland's Tony Wendice, in *Dial M for Murder*, uses the glass phone booth in the lobby of a restaurant to call his wife and thereby put her in position to be strangled by a man hiding behind a curtain in their apartment. In

The Sting, which takes place in 1936, Johnny Hooker and Henry Gondorff scam the mafia boss Doyle Lonnegan out of $500,000 by placing a series of false bets on a horse race from a restaurant phone booth. Paulie, the criminal leader of Henry Hill's gang in Martin Scorsese's *Goodfellas*, hates phones, won't keep one in his house, and so receives all his calls "second-hand," then has an associate return the call from a public phone. When Benjamin Braddock finally asks Mrs. Robinson to have a drink with him, he does so from a glass phone booth in the lobby of the Taft Hotel. After she arrives, he later calls her in the bar from the same phone booth in the lobby, letting her know that he has procured a room.

Unlike most "real" phone booths and pay phones, none of these phone booths smell rank and all of them are operable.

FIGURE 18 *Dial M for Murder*, dir. Alfred Hitchcock, 1954.

The most interference that occurs in the examples cited above is that Tony Wendice needs to wait for someone to vacate the phone booth before he can use it. But as Renata Adler notes, phone booths in cinema often made people more vulnerable than safe: "In almost every thriller a point is reached when someone, usually calling from a phone booth, telephones with a vital piece of information, which he cannot divulge by phone. By the time the hero arrives at the place where they had arranged to meet, the caller is dead, or too near death to tell."[15] The phone booth outside of the gas station that Detective Arbogast steps into, and then does not leave, comes to mind, as does the phone booth at the beginning of the 1932 *Scarface*, in which the head of the mafia, Louis Costillo, is shot.

Phone booths seem to have facilitated just as much crime in real life as they did in fiction. In 1921 police engaged in a "telephone chase," following a man who was hopping from phone booth to phone booth, asking for ten thousand dollars in exchange for a high-society woman who had gone missing. In 1932 a man was sentenced to three months in prison for cutting 627 wires in various telephone booths because he was unhappy with how they performed. The late twentieth century saw another version of the *Telephone Booth Indian*: drug dealers who found the anonymity of pay phones invaluable. In July 1984, a pipe bomb exploded in a Minneapolis phone booth. In September 1997, a man armed with a razor raped a woman in a Park Avenue booth as pedestrians walked by. As early as 1994 there were only twelve phone booths remaining in New York City, but residents were campaigning to have them turned into kiosks, reporting that they functioned more as meeting places for prostitutes and homeless shelters than they did as housing for telephones. An unmarked phone booth, by contrast, implied an abiding innocence about the area in which it was situated. In an article published in 1969, Massena, New York, was preparing for the President to visit: "Signs of the good life were all about them here: clean streets, undefiled phone booths, and bunting-draped stores farther downtown where all has been made ready for the visit tomorrow of President Nixon."[16]

The lives of others

Personal and public phones continue to be embroiled in controversies surrounding privacy. In 2005 the *New York Times* revealed that, in 2002, President George W. Bush had authorized the warrantless wiretapping of thousands of phone calls from specific targets; because of technical malfunctions, however, the phone calls of some non-targeted people had also been collected. In June of 2013 Edward Snowden disclosed that the NSA had been collecting phone records for all domestic calls using Verizon, Sprint, and AT&T networks. Although the court order authorizing and commanding this surveillance emphasizes that what is being monitored is not the content of these calls but their origin and destination, length, and frequency, people were rightfully upset, and surprised. Earlier iterations of the surveillance mechanisms used were first implemented by AT&T in the sixties and seventies, when they began monitoring their lines in order to find and prosecute the "phone phreakers" who had figured out an ingenious way to make free calls on both private and public phones.

In October 2014 a story broke revealing that the advertising firm Titan had installed hundreds of "beacons" in public phone kiosks across the city. Beacon technology employs Bluetooth to send signals to mobile phones with "receptive apps" that can then allow ads to be sent or the phone owner's location to be monitored. New York's Department of Information Technology sanctioned the installation of beacon technology without informing the public, and BuzzFeed was quick to note the more sinister implications of this technology: "The spread of beacon technology to public spaces could turn any city into a giant matrix of hidden commercialization—and vastly deepen the network of surveillance that has already grown out of technologies ranging from security cameras to cell phone towers."[17] Although Titan emphasized that none of their beacons was being used for surveillance or to collect data, and that the technology only worked successfully if people had specific apps on their phone, the immediate public outcry revealed the extent to which we feel ourselves controlled by both social media and advertising, and how the two work more and more in concert with one another. The beacons were soon removed.

9 GLASS CASE OF EMOTION

I am always surprised to never encounter a phone booth in an Edward Hopper painting, to not see the fragment of one, say, in the corner of the diner in *Nighthawks*, or in the background of *Automat*, or outside the restaurant in *Chop Suey*, because phone booths are public spaces but they tell us a lot about personal isolation, what Hopper captured so well in his depiction of urban environments and even, somehow, in the slant of sunlight against a wall. No, it was literature and film that first helped us see the psychological complexity of such a structurally simple space, where we could feel the limit of our own existence. Not only did pay phones advance the plot—Phyllis Dietrichson and Walter Neff in Billy Wilder's *Double Indemnity* would never have gotten very far without being able to collude with each other over pay phones—the structure of the booth framed numerous dramatic encounters, with others and, moreover, with the self alone. "A man is talking on the telephone behind a glass partition," writes Albert Camus. "You cannot hear him, but you see his incomprehensible dumb show: you wonder why he is alive."[1]

Phone booths often showed communication for what it is, or can sometimes be: a pantomime of connection. In his drawling, atmospheric thriller, *The Killing of a Chinese Bookie*, Ben Gazzara's Cosmo Vitelli stands in a well-lit phone booth off a freeway exit in Los Angeles, talking to the manager of his strip club. Having celebrated making the final payment on a longstanding debt by immediately losing twenty-three thousand dollars to a bunch of gangsters in a poker match, Vitelli has no other way to absolve his new debt than to make a hit on a the head of a rival mobster. When he steps into the phone booth Cosmo is on his way to kill the bookie in question, but he nevertheless attempts to maintain the veneer of easy sophistication when his car stalls out and he must wait for a cab. When he calls his manager, Vitelli is killing time but also attempting to extend it. He tries to find out what numbers are being performed, chides his manager for his ignorance, and moves restlessly around the booth until he finally breaks out into song—a delirious but sincere rendition of "I Can't Give You Anything But Love." The camera is always close to Cosmo's face and initially we observe him as though we are outside the phone booth, waiting to use it, but about halfway through the scene the frame of the booth dissolves and we are inside the cramped space, cheek to cheek. It hurts to be this close, because Cosmo is essentially a good, if unlucky, man, and in this moment he is desperately trying to maintain appearances, to his manager yes, but most keenly to himself.

FIGURE 19 *The Killing of a Chinese Bookie*, dir. John Cassavetes, 1976.

In Haruki Murakami's novel *Sputnik Sweetheart*, the main character, Sumire, loves to think about Laika, the dog that the former Soviet Union sent into space aboard Sputnik II and was ultimately lost. She imagines the small dog looking through the even smaller window. "In the infinite loneliness of space, what could the dog possibly be looking at?"[2] In the dog's hapless yet observant wandering is a portrait of Sumire herself, who has dropped out of college in Tokyo to be able to write. In order to call anyone she must walk to a phone booth a few hundred yards from her apartment, something she does quite often, in the middle of the night, and seemingly only to call her friend K, who narrates the story and who is also, as it happens, in love with her.

The phone booth is described as average—dirty, plastered with ads—the moon above it as "orphaned," but it is a place in which Sumire feels comfortable sharing her most disturbing questions and thoughts, those questions and thoughts which cannot wait for an answer or, at the very least, an opportunity to be voiced. K never fails to take the call, even at three in the morning, but from the outset of the novel we know that Sumire falls in love with a woman who is seventeen years older and that K is fated only to orbit Sumire, perpetually circling but never intersecting. For a long time Sumire, too, seems destined only to orbit her love, Miu, who does not or cannot return her feelings. In her effort to refashion herself into someone she thinks Miu could love—she stops writing, quits smoking, changes her wardrobe—Sumire feels as if she is living in a universe in which she has lost everything, even gravity. As her sense of self dissolves, she is increasingly lonely and alienated—a satellite wandering through the "unimpeded darkness" of space—and her phone calls to K increase until Sumire first moves and then mysteriously leaves on an impromptu trip to Europe with Miu. It's during the course of that trip, when they are in Greece, that she disappears completely.

One of Sumire's central concerns is how to reconcile what we know and what we don't know. "Inside of us what we know and what we don't know share the same abode. For convenience' sake most people erect a wall between them. It makes life easier. But I just swept that wall away."[3] How?, one might well ask, and Sumire's answer is not thinking but, instead, dreaming. You accept things that are impossible to understand and leave them as such, which is what both Miu and K must do when they cannot find any trace of Sumire, and perhaps what Sumire has done in disappearing. After traveling to Greece in order to find her, K muses that she must have found an exit from this world, where she couldn't be with Miu, and then found a world where she could. But one day K does hear from her. "Hey, I'm back," she says, calling from "our good old faithful phone booth. This crummy little square box plastered inside with ads for phony loan companies and escort services."[4] Though she seems to return to the phone booth she once frequented on a regular basis, she has no larger idea where she is. She promises to get change and call him back with more details, but she never does. K looks out the window and feels comforted that they are both standing beneath the same moon, and, moreover, by the idea that there is no rush.

The novel focuses on Sumire, but it is also, indirectly, about the narrator, who describes himself as having drawn an "invisible boundary" between himself and other people. "As you might guess," he says, "I led a lonely life." Later he finds teaching, that turns out to be a natural fit for him and alleviates some of his loneliness, but still, it is with Sumire that he engages with the big questions, the questions religion and philosophy give us the language and excuse to ask: "Who am I? What am I searching for? Where am I headed?" When he travels to Greece to look for her, he sits in the airport and feels that he and Sumire and share an unstable shifting world: "We were nearly boundless zeros, just pitiful little beings swept from one kind of oblivion to another."[5]

Murakami's fictional universes are full of gaps and inconsistencies, places where planks in reason break and logic gives out completely. In such a context, where people are perpetually revolving around each other through space, the phone booth is appealingly still and constant. Sumire might have found an ultimate way to escape from a painful, contradictory world, but the inklings of a healthier exit appear in the phone booth, when she reaches past the orbit of her own solitude and attempts to connect with another. "And I really wanted to see you, too," Sumire says to K in their last conversation. "When I couldn't see you anymore, I realized that. It was as clear as if the planets all of a sudden lined up in a row in front of me."[6]

In order to escape numerous men trying to collect on the money they think her dead husband has stolen from them, Audrey Hepburn's Reggie Lampert in the film *Charade* takes refuge in a phone booth. Why is it that people in movies always hide in phone booths when everyone knows phone booths are the very last places they should go? After one of her pursuers forces his way into the booth and flicks successive lit matches on her dress, Regina cowers lower and lower, hurriedly brushing off the flames. What was meant to be a haven is now the perfect incineration chamber. A few minutes later Peter Joshua (Cary Grant) makes a belated entrance and asks her what she is doing there, to which she responds, "I am having a nervous breakdown."

Phone booths are good places for nervous breakdowns. *Charade* was released in 1963, the same year as Hitchcock's *The Birds*. When the deadly birds resume their attack on Bodega Bay, Melanie Daniels, played by Tippi Hedren, leaves the diner from which she had been watching the maelstrom and runs outside to a phone booth. From a viewer's perspective her decision is inexplicable—there is a working phone in the diner—but Hitchcock's reason is clear. The scholar David Trotter explicates Hitchcock's choice quite succinctly when he writes, "What he gets from Melanie's mistake is an image of isolation and exposure, as she twists and turns in torment in her transparent cubicle, and the glass

FIGURE 20 *The Birds*, dir. Alfred Hitchcock, 1963.

shatters."[7] What drives Melanie into the telephone booth, and what cannot get her out, is a more acute threat than the birds.

The idea of the phone booth as a perfect container for and representation of the unconscious, as Trotter points out in the same essay, is made explicit in Joel Schumacher's 2002 film *Phone Booth*. In a glass phone booth on the 53rd and 8th, "the last vestige of privacy in Manhattan's west side," which is scheduled to be replaced by a kiosk the next day, a philandering publicist is held captive by a sniper's rifle until he is able to admit his transgressions. A boy from the Bronx intent on recreating himself, manipulates everyone trusting enough to believe him, falsely promising to turn people into gods in order to shore up his own uncertain identity. When he picks up a ringing pay phone he begins to engage with someone who perceives himself to be his priest and the phone booth becomes his confessional. His punishment is to act out the scene of confession publicly, at gunpoint.

A short Spanish film from 1972, *La Cabina*, makes Camus' image of the dumb show quite literal, and takes the conceit in Joel Shumacher's *Phone Booth* a step further. After dropping his young son off at the bus stop, a middle-aged man—balding, average weight—steps into a newly installed phone booth in an empty square to make a phone call. He shuts the glass door behind him, only to find that the phone doesn't work and that when he tries to leave, he can't: the door won't budge. No matter how much he pushes from inside and, eventually, how much people push from outside, he remains stuck. Almost immediately, spectators begin to gather and gawk. Kids make fun of him. Rather than empathy, the situation inspires discomfort and scorn. The man inside the booth cannot communicate in any meaningful way with people outside, nor can he call someone for help. When the telephone company finally arrives and takes away the booth with the man inside, the people cheer. Like Joseph K and also K, this anonymous man is caught in an inexplicable situation through no fault of his own. The employees of the telephone company, who remain largely invisible in the truck, take him to a futuristic building where he sees the remains of others like him, all in various states of decay, locked in their respective glass coffins.

Transformations

You enter a space in one way and leave it in another. Nick Carraway abandons the comfortable familiarity of the Midwest—"the ragged edge of the universe"—because he hungers for something more central, only to return a year later, considerably chastened, wanting "no more riotous excursions with privileged glimpses into the human heart."[8] This provisional space of the phone booth has often been the backdrop for and agent of acute psychological transformation. In 1957 George Langelaan published a short story called "The Fly" that David Cronenberg later made into a movie, in which Seth Brundle, a brilliant young scientist, discovers how to transport materials from one place to another without causing any destruction. The process, which he calls "disintegration-reintegration," uses two repurposed phone booths he calls Telepods. One evening, he decides to teleport himself, unaware that there is a small fly in the pod with him. Although the experiment first appears to be successful, he gradually begins to transform into the fly, shedding his human form and his humanity in successive layers. "Am I becoming a hundred-and-eighty-five pound fly?" asks Brundle. "No, I'm becoming something that never existed before. I'm becoming Brundlefly."[9]

One of the most seminal scenes in *The Godfather*, released in 1972, occurs when Michael, the youngest Corleone, finds out that his father has been injured. He gets the news in a phone booth, still dressed in his military uniform, right after he has been shopping with his girlfriend Kay, played by Diane Keaton. The news Michael receives in the phone booth effectively obliterates any chance he might have had to start a new life with her distinct from that of his family. Now that his father has been hurt, possibly killed, he will have to step up. As we watch Michael's face transform when he hears the news, Kay becomes visible in the background, outside the phone booth. No matter how close she comes to Michael, she remains an observer, outside of the phone booth and the life that is unfolding for him within it.

"This definitely isn't the most comfortable place in the world to switch garments," states Clark Kent after explaining why he is leaving to Lois, "but I've got to change identities—and in a hurry!" Undoubtedly the most iconic transformation that occurs in a phone booth is when Kent, an awkward, diffident journalist for the *Daily Planet*, periodically changes into the ever-capable, heroic Superman in various phone booths throughout Metropolis. Superman's appearance always amazes everyone around him, because he seems to materialize out of thin air. Similarly, the public knows nothing of his origin story, guarded by the Fortress of Solitude as successfully as the phone booth hides Clark Kent's clothes. When Jerry Siegel and Joe Shuster sold *Superman* to DC Comics in 1938 the Second World War was about to begin, Stalin had decimated his population, and many Americans were still faring poorly because of the Depression. Although Superman entered a world that was already populated with heroes—Shadow, Hugh Danner, and Doc Savage among them—it was Superman who really captured the national imagination. Comic books outsold all other reading material among soldiers serving in the Second World War, and *Superman* was widely considered to be the most popular comic.

While the phone booths Superman changes in are as iconic as the character himself, in the comic strips, television series, and films, Clark Kent uses the phone booth only sporadically; it was much more common for him to change his clothes in the storeroom of the *Daily Planet*. "The first time Clark Kent ducked into a phone box to change into Superman was in November 1941 in *The Mechanical Monsters*, the second of the animation films," writes Larry

Tye. "It proved convenient enough that he did it again on the radio, in the newspaper strips and comic books, on Broadway, and, most famously, in the movies."[10] The older, wooden phone booths offered privacy in the way more modern glass booths could not. While phone booths are noticeably missing from almost all of the movie franchise, in the 1978 film, starring Christopher Reeve, Clark is looking for a place to change, sees a phone kiosk, and visibly dismisses it, highlighting what an insufficient refuge a kiosk is in comparison to a booth.

Although Superman did not make regular use of phone booths, the two are inextricably wedded in the popular imagination, and one reason might be related to how phone booths were portrayed in the first half of the twentieth century. Lighthouses for the road, inanimate soldiers "on-call" twenty-four hours a day, rain or shine, phone booths were promoted as a panacea for every difficult or inconvenient situation. As such, they were a fundamental element of the utopic society we have long hoped might be achieved through technology. These structures offered unprecedented convenience and became precursors for much of today's technology, which is not only responsive to our needs but also virtually indistinguishable from our bodies. In 1960, for instance, the Westinghouse Corporation designed a prototype for a system whereby someone could leave their house, having turned all the switches to automatic, and then call home from a pay phone and control the appliances from wherever he or she was.

The ghost of this relationship remains in how pay phones often still work in situations when cell phones go dead. On the morning of September 11, for example, most of New York City was a cellular dead zone, but landlines still worked. Ironically, after September 11, pay phones with booths were often targeted for removal—as in the bank of sixteen phone booths in the Western Union Building at 60 Hudson Street. A "nerve center for several telecom companies," the building became a potential terrorist threat, and while the phone booths had once offered valuable privacy and insulation, they now offered, in the public imagination at least, a perfect place for a terrorist to hide a bomb. A security guard needed to be dispatched every time someone needed to make a telephone call, which quickly resulted in the phones becoming off-limits for public use.[11]

"If there is an infinite aspect to space," writes the Russian poet Joseph Brodsky, "it is not its expansion but its reduction. If only because the reduction of space, oddly enough, is always more coherent. It's better structured and has more names: a cell, a closet, a grave."[12] Though the phone booth's dimensions are practical—perfectly designed to fit one person comfortably for a temporary period of time—there is something also poignant about these measurements, something that drives home our essential isolation from each other, what Augustine thought was a product of our fallen nature and what John Locke, writing hundreds of years later, perceived as inherent to our existence.

On the northwest coast of Japan, the tall cliffs of Tojimbo drop two hundred feet into the sea. These cliffs are one of the most popular places to commit suicide in a country with the world's highest suicide rate. In addition to having policeman regularly patrol the cliffs for potential suicide victims, city officials have installed outdoor lighting along with two pay phones and spare change so that people might call the national suicide hotline. The phone is bright green and sits in a glass booth that affords a view of the ocean.

10 THE GOD BOOTH

Approached from the west, the Mojave Desert officially begins on the eastern flanks of the San Gabriel, San Bernardino, and Tehachapi mountain ranges, along the fractures of the San Andreas and Garlock fault lines. Unofficially it starts where the Joshua trees begin to grow. Bordered to the north by the Great Basin and to the south and east by the Sonoran Desert, the Mojave spans fifty-four thousand miles and four states; a basin and range topography, it contains the lowest point in the country as well as peaks surpassing ten thousand feet, not to mention Las Vegas, a city famous for its elevations and depressions.

In July of 1999, the Holy Spirit directed Rick Carr, a fifty-one-year-old Texan, to travel a few hundred miles from his home to answer the calls made to a phone booth in the middle of this desert.[1] Located near the California-Nevada border at the intersection of two dirt roads, seventy-five miles southwest of Vegas, the phone booth shared property with desert tortoises, saguaro cactii, and sagebrush. It is a landscape of asceticism and religious vision, at whose edges the American military hovers—there are seventeen United

States military sites scattered throughout the Mojave, one of which is the largest Marine Corps base in the world.

Carr did not take this decree lightly: he camped beside the booth on the desert playa in scorching heat for thirty-two days. During that time he answered over five hundred calls, many of which came from someone named Sergeant Zeno, who said he was phoning from the Pentagon.[2] What was there, in the middle of the Mojave, was a ghost of what *had* been there: a phone booth positioned along phone lines stretching from central Washington to Southern California that had been installed during the Second World War and would, the government and telephone companies believed, be immune to a potential attack from the Japanese. The phone booth was installed at the civilian request of Emerson Ray, on behalf of the local volcanic cinder miners who would be well served by having access to a public phone. Initially a hand-cranked magneto, it eventually became a coin-operated pay phone, first equipped with a rotary dial, then a touchtone.

Carr is only one of many pilgrims who have trekked to the Mojave Phone Booth, and Sergeant Zeno only one of many callers. A Los Angeles resident was compelled to make the trip simply by seeing the image of an isolated telephone icon on a map of the Mojave, and it was his description of the excursion published in an underground zine that inspired Godfrey Daniels, who was also a fan of the band Girl Trouble, whose zine it was, to call the phone in the hopes someone would pick up. Trying to explain his fascination with the phone booth, Daniels (known as Doc to his friends) said, "I don't know if in the age of cell phones it's the same, but in those days when you were out in the desert, you were on your own. You couldn't call people." A phone booth in the middle of this no-man's-land, then, was "kind of like if somebody was on the moon, and you could talk to somebody on the moon."[3]

Daniels was "prepared to call for years" until someone picked up, but in less than a month Lorene Aiken, a local resident who worked at the nearby cinder mines, heard the ring and answered. Aiken had never heard the phone ring before, but she chatted amiably for a few minutes with Daniels, who decided to actually make a trip out to the phone booth on his way to Burning Man, the annual techno-hippie festival that takes place in the Black Rock Desert of Nevada. After his return, Daniels, a computer programmer by trade, made a website devoted to the phone booth: "I thought that was about as far as it would go. Next thing you know, I'd go to my P.O. box, and there would be clippings about the Mojave Phone Booth from newspapers in languages I didn't know how to read. It just spread."[4] One of the first Internet "memes," before the concept of memes existed, the Mojave Phone Booth soon began appearing in newspapers and magazines throughout the world, and the phone, whose number Daniels listed, began ringing constantly.

The emptier the space, the more likely we are to interpret and find meaning in what is there. Everything seems to evoke something else, a geologic or social or personal history that's become extinct and left clues as to its existence in the current version of the landscape holding sway. By the time Rick Carr arrived, the walls were pocked with bullet holes and the glass casing had been shattered, leaving only an aluminum skeleton lined with candles, license plates, rosaries, and other votaries. Between 1997 and 2000, when Pacific Bell retired the number, the phone received thousands of calls, dozens each day. It also received thousands of visitors, devotees as zealous as adherents of any religious sect, who came from all over the country and abroad to answer the phone and to record the conversation in a log, a handwritten hybrid of a phone book and visitor's diary.

"Wait till the sun goes down," says one visitor to the booth, "because there's nothing more surreal than a lone phone booth ringing with not an ear to hear it. It's very *Twilight Zone*, you know what I mean? That's the fun part of it. It's just a stinkin' phone booth in the middle of stinkin' nowhere." As another person put it, "It's more than real. It's reality!"[5] One woman plaintively commented that she had grown up in the area and remembered the phone booth "when it was just a phone booth." But by that time the phone booth had long stopped being just a phone booth and was, as a journalist for the *Orlando Sentinel* wrote, a "pay-phone Woodstock." When asked why they phoned, most of the callers' answers could be distilled to this: because there was a chance someone would pick up. The telephone affirmed the magical possibility that our lives are mysteriously linked, that the most ordinary events glitter with unseen significance.

And then, one day, the commune was disbanded, the connection cut, swept away as quickly as the scant amount of rain that falls on the desert each year. Unbeknownst to Daniels, the area in which the Mojave Phone Booth was located had recently been designated a nature preserve, and the large amount of foot traffic, which was without precedent, was damaging the delicate ecosystem. The National Forest Service demanded that the phone be removed, and in May of 2000 it was.

When telephone booths disappear they usually do so gradually, in serial stages of decay, abandonment, desecration, and, finally, removal. In this case the booth vanished in a few hours, at the hands of a couple of men, leaving only the plinth on which it rested, which itself became an altar where people left various offerings. In another few months the forest service carted that away too.

One woman asks if she might trade a bit of Salem graveyard dirt for a shard of the Mojave Phone Booth, but the vast majority of the texts on Doc's website are accounts of people's visits to the booth, or getting lost in their attempt to find the booth. That the Internet made the phone booth famous is, of course, ironic—the chief cause of the booth's obsolescence being the agent of its renewal.

FIGURE 21 The Mojave Phone Booth. Photo © Doc Daniels, used with permission.

Burning Man

In some ways, there is little to distinguish between calling a phone booth in the middle of the desert that may or may not be occupied by someone who could—or even would—answer, and addressing an equally hypothetical god in praise or supplication. Both involve a leap of faith, however seriously or insouciantly that leap is made, and both are predicated on a future that has not been circumscribed by the limitations of the present. One could say, of course, that the presence or absence of belief distinguishes these acts, but in the instance of the Mojave Phone Booth, the booth itself had become an object of devotion, a reliquary of isolation, and the need for connection. Thomas Watson and Alexander Graham Bell rather strategically chose to demonstrate how the telephone worked in churches between sermons and hymn singing. The people who called the Mojave Phone Booth and the people who picked up were people who believed, consciously or unconsciously, in technology's ability to connect us in meaningful ways.

Burning Man, the temporary autonomous zone that arises and disappears within a week every year in the Black Rock Desert, a few hundred miles northeast of where the Mojave Phone Booth once was, and where Godfrey Daniels ended up after visiting the Mojave Phone Booth for the first time, is premised on this belief. Originally taking place on Baker Beach, in San Francisco, where a few friends got together annually on the summer solstice, the event grew exponentially and eventually moved to the desert. Having undergone several permutations but always touted as an experiment in "radical self-expression," the heart of the festival continues to be what people or groups of people make and then either give away or destroy. Over the years, during which Burning Man has achieved the status of a permanent business and a temporary city—instituting laws, requiring permits, and even building along a grid structure—the festival has a become more and more technically sophisticated. Unsurprisingly, perhaps, a significant number of "Burners" are people in the tech industry, the traveling utopic impulse that had once grounded itself in a return to the land finding a contemporary articulation in technology that ostensibly seeks to unite people in a web of shared information.

As Erik Davis details in *Techgnosis*, his pre-millennial study of technology's spiritual underpinnings, the attempt to harness technology as a means of achieving our human potential goes back to the image of New Jerusalem descending at the end of Book of Revelation: "As a futuristic image of heaven on earth, the New Jerusalem would directly inspire the secular offspring of Christianity's millennialist drive: the myth of progress, which holds that through the ministrations of reason, science, and technology, we can perfect ourselves and our societies."[6] In this new city described in Revelations, "prepared as a new bride beautifully dressed for her husband," there is neither mourning nor death, "for the old order of things has passed away." And indeed, Burning Man is nothing if not an attempt to create and live in the perfect society. Aside from tea and coffee, nothing is sold. Instead, people give and receive gifts. The festival's Ten Principles, which emphasize generosity and responsibility, are touted as a code of conduct. People are encouraged to experiment with various ways of "opening and expanding" the self, so long as, by the time they leave, they have left no trace. As a natural consequence of its openness, perhaps, Burning Man has also become a place to test out wild ideas with impunity, including beta versions of both Google maps and Tesla cars.

For all of the ways the festival has changed in response to the digitization of humanity, however, communicating with God remains an analog affair. All one has to do is find the God booth, a repurposed glass phone booth placed on an esplanade somewhere in the midst of what becomes, momentarily, the second-largest city in Nevada, and call the listed number. A perfectly flat, dry lakebed, where a jet car broke the sound barrier and the first amateur rocket rose high enough to be deemed space flight, the Black Rock Desert is a place ripe for religious encounter. What the conversation consists of depends on the caller, and the God one encounters depends on the time one calls. I read a couple of accounts and was surprised to find them so positive, so devoid of snark. One person reported that God (at the time a woman with a soothing voice) gave her some very useful advice before reciting the Serenity Prayer, then talked to the next person in line about fantasy football strategy for the upcoming season. Another man reported that his God had a Middle-Eastern accent and apologized for it before sharing some wisdom about relationships.

No matter the intentions with which it was constructed, or used, the God booth highlights the idea that prayer, in its various forms across various religions, is a *call* to the divine, prayer being the technology that is intended to overcome all time and distance, what early advertisements suggested was the telephone's chief ability. (Oliver Stone depicted Andy Warhol as having a "god phone," which gave him a direct line to the divine, although he purported to have nothing to say to God.) There is a distant echo in the way the telephone lines that preceded fiber optics depended upon establishing a secure circuit by which to convert electric signals into sound and the incense one might burn in a Daoist temple before prayer so as to let the smoke clear a channel of communication.

The God Booth's counterpart at Burning Man was a pay phone that really did depend on secure circuits. In the early 2000s, a Burner installed a working pay phone for other attendees to stay in touch with the world they had temporarily left. "Ideally," stated the creator of the pay phone, "I wanted a traditional 'Superman' style booth, and those can be found, but cost a fortune to ship, so we went with a more modern pedestal style phone. The goal was to have the phone just sitting there, mounted on the desert floor, connected to nothing, yet working, just where it shouldn't. We did it, and the results were amazing and surprisingly emotional. People refused to believe it, and cried out with joy when it became real. In spite of problems, about 1,600 calls were made all over the world."[7]

New Haven

During the winter and spring of 1999, a few months before Rick Carr made his pilgrimage to the desert and a few months before I left for Ireland, I would only talk to people from a specific phone booth in New Haven, Connecticut. The booth was on the corner of Elm Street and College Street, adjacent to the city green, a few blocks from the apartment I shared with three friends; it was a glass box with a half-eaten directory dangling on a chain from the shelf beneath the telephone. My last semester of university consisted of a slow oscillation between working on my thesis and talking on that phone.

"You could say anything, confess, if you desired," writes David Lazar about his fascination with pay phones, "your most heinous crimes, most lingering guilts, without the danger of being traced."[8] Because of their anonymity, because of their privacy, public phones have always been associated with intimate disclosures. The structure of the booth itself recalls the physical presence of the confessional, which exists in the popular consciousness as the habitation of truth and absolution.

Confession, in which one may confess one's sins, either grave or venal, and be absolved by a priest acting as a "minister of God's mercy," is as old as the church itself, but it was not made a requirement until 1215, by the Fourth Lateran Council. Until the seventh century, confession occurred mainly in monasteries and was reserved for the clergy. If a layman wanted to confess, he or she had to do so publicly, and the punishments tended to be quite harsh—years of imprisonment and abstinence. Beginning in the sixth century confession, however, became both more accessible and more private, as Irish clergy developed *penitentiales*, "manuals which listed and classified sins as well as punishments to be imposed for each specific sin."[9] Still, the punishments remained extreme, involving years of asceticism and deprivation. When confession became mandatory, the manuals were largely forgotten and punishment was determined on an individual basis by the clergy.

As the apparatus and logistics of confession changed, so too did the nature of confession, which became less concerned with the acts themselves than the intentions with which those acts were committed. As the soul became an increasingly private place, so too did the location in which confession occurred. After the Counter-Reformation, confession often took place in the booths we are familiar with today and became a reconciliation between the individual and God, rather than the individual with the church. The confessional was designed to facilitate this anonymity and also to prevent the penitent and the confessor from coming into physical contact with each other and thereby incur more sin. But the development of the confessional also marked a corresponding development in confession, whereby sin and absolution became "less a matter of submission to God's law, and more a question of the sinner's relation to his or her own conscience, albeit a conscience informed by that law."[10] The confessional box, which divided the sinner from the priest, represented and facilitated that shift.

Not only did the confessional box reflect the changing nature of confession, it also suggests a consideration of how public revelations of private misdeeds might affect the well being of a community. A confessional distinguishes itself from the church in which it is situated just as, for the believer, a church is a different kind of space from the street in which it exists. The door of the church is a threshold that not only marks the difference but also acts as a passage between the two worlds. Not only did the closed confessional preserve the distinction between the public and private spheres, the closed confessional also preserved the penitent from disturbance, thus facilitating the focus of attention on the act of contrition.

Outside of the church, various species of confession occur in therapy, autobiography, and in conversation, but it remains, in the words of Foucault, "a ritual that unfolds within a power relationship, for one does not confess without the presence (or virtual presence) of a partner who is not simply the interlocutor but the authority who requires the confession, prescribes and appreciates it, and intervenes in order to judge, punish, forgive, console, and reconcile. . . ."[11] For me, confession of a kind occurred in a dirty phone booth in New Haven, a city I had not found to be a haven at all. The phone booth preserved the anonymity of my disclosures, even when I was speaking to the people who knew me best. The invisibility, however nominal, is what made the admissions possible. The space simultaneously consecrated the exchange and maintained my distance from everything that had driven me to it: opportunities lost, failures sustained—the accumulation of the person I had somehow come to be.

"For the religious man space is not homogenous," writes Mircea Eliade.[12] Certain spaces are sacred, characterized by their significance and structure in comparison to the triviality and formlessness of everything that surrounds them. "When the sacred manifests itself," Eliade goes on to say, "there is not only a break in the homogeneity of space; there is also a revelation of an absolute reality."[13] The manifestation suggests a larger reality from which one can orient and create the rest of the world. Celtic Christians referred to "thin" places, where the distance between heaven and earth diminished. In such places they were able to communicate more easily with the gods.

"Within the sacred precincts," continues Eliade, "the profane world is transcended."[14] Whether made of wood or glass, the phone booth stands apart, and is made to stand apart, from the normal flow of life in which it is situated. Although primarily functional, its existence suggests something more profound: the necessity of sanctuary, without which life is untenable.

Desert solitaire

The Mojave Phone Booth's main competitor for fame was another pay phone thirty miles east, in Lanfair. Like the Mojave Phone Booth, it was installed for the use of cattle ranchers and the few other residents of the area in the 1960s. After repeatedly being vandalized the booth was moved to a location closer to the Omni navigation system used for orientation purposes by passing aircraft, and was downgraded to a kiosk in the eighties that was removed in the early aughts. It enjoyed some attention—some visitors, a few newspaper articles, and a mention on Doc's website—but examining the respective fates of the Mojave and Lanfair phone booths is a little like comparing the trajectories of Bob Dylan and Dave Van Ronk. One of them became world famous, was deified, and made into an oracle, and the other remained just a phone. For a time people periodically come to attach random desk phones to the pole, either their own or ones you might find in a thrift shop, in memorial.

"It was the blank space on the map surrounding that phone that intrigued me," Daniels said, explaining why he had been compelled by the Mojave Phone Booth and not much inclined to visit the one in Lanfair. "There was a lot more stuff around that one." Located in the southwestern part of the Mojave, also near the California-Nevada border, Lanfair Valley is a ghost town. The structures there are ruins, and so don't turn up on a map. When Los Angeles was beginning to roil into the entertainment and housing complex it later became, small groups of people were retreating to the Mojave to establish Socialist enclaves. It helped that in the 1900s, the Mojave was also enjoying one of its few wetter spells, making the area seem more inviting to farmers and ranchers. Lanfair was one of the only areas whose democratic reach extended to race, and for some time the area was heavily settled by African-American families. At one point an orphanage for

African-American children was established but quickly failed. Homesteading began here in 1910 and was over by the 1920s. According to the desert historian Dennis Casebier, "most of their improvements were recycled or burned by the big cattle company."[15] Sitting at a little over four thousand feet in elevation, the area contains the Grotto Hills and the Longhair Buttes, as well as, at the intersection of Cedar Canyon and Longhair roads, the ruins of what had been Lanfair—a school, post office, and store.

The desert is rarely friendly to ambition, and the fate of Lanfair is a testament to that. In a context like this, the phone was just another ruin. There's more melancholy here than mystery and people's attempts to see it seemed to trail off, just as the attempts to live in Lanfair trailed off and ultimately gave out. It's *this* phone booth, sometimes a phone nailed to a pole, sometimes just nicks indicating where the nails once were, that is the true oracle of the desert.

11 ONLY CONNECT

At the beginning of a class day in early March of 2011, a cluster of students huddled around my laptop computer watching YouTube footage of cars, houses, trees, and people being swept away by the tsunami battering the east coast of Japan. The water was speeding, defying all law and order, taking what it wanted and submerging the rest. We stared at the water coursing through the streets that were no longer streets, at the town that was, momentarily, no longer a town but a part of the ocean. We had woken up to the tragedy while it had struck Japan in the middle of a weekday afternoon, when people were in the dead center of their lives. Even after I shut my computer off and took out the book we were in the middle of studying, signaling that I was ready to begin class in earnest, the students continued to steal furtive glances at the footage on their iPhones. They were concerned but also fascinated, ashamed to be compelled by real tragedy but unable to look away.

For my students, and almost everyone younger than thirty, pay phones and phone booths are ancient ruins, relics of an earlier age that seems so entirely different from how we live now that they make thirty years ago feels like the distant past. Small children are more attuned to the potential of a phone in a box than teenagers who, if they see pay phones and phone booths at all, tend to dismiss them as dysfunctional occupiers of public space. Young children can still find them magical, a world within a world with only one belonging—a telephone. Surely that telephone can reach more people and more places than the mobiles that belong to everyone and are always in use. Surely that telephone is special, and it is. One woman in her forties described to me how her mother had refurbished an old wooden booth and installed in the corner of an unused room in their house. The booth was empty but it was where the daughter and her friends went to exchange of all their secrets, and where she eventually retreated to experience her most intense emotions. Even pay phones removed from their usual context are alluring. Another friend lived in a house whose previous owner installed an antique pay phone, and which her father kept as a work line and she and her friends spent endless hours on, depositing coins and "playing operator."

Any recent article about the current state or the future of pay phones will cite numerous statistics to prove that millions of people still find a practical use for them, and that in developing countries they continue to be installed more rapidly than they are removed. The statistics go something like this: approximately fourteen million households in America do not possess cell phones and one hundred and forty million people do not possess landlines. Even with the dramatic decrease in working pay phones—according to the American Public Communications Council (APCC), there are fewer than five hundred thousand pay phones remaining in the United States—they continue to make millions of calls each year, mostly by people who cannot afford mobile phones or landlines, but also by those in more remote areas where cell service is still not ubiquitous.[1]

These statistics attest to a persistent need for universally accessible, cheap communication. Many journalists have coyly used the metaphor of an endangered species to discuss the disappearance of the public phone. Even if tongue-in-cheek, this language suggests that phone booths and pay phones were integral elements of urban and rural systems, and that their absence will have significant cascade effects. Junichiro Tanizaki, in his essay "In Praise of Shadows," reflects on how the fountain pen, "an insignificant little piece of writing equipment, when one thinks of it, has had a vast, almost boundless, influence on our culture."[2] One could make a beautiful case for the boundless influence of any of our undersung objects, and books and films, as well as our own memories, have been making that case for pay phones and phone booths ever since they were invented. Just as the invention of the clock changed our subjective experience of time, public phones transformed our subjective experience of communication. They made communication more universal, yes, but they also made it more subject to accident and coincidence. They afforded us some privacy, but also exposed us to the rest of the world. They gave us more control but also took a bit of control away, just as mobile phones do now when they "go dead," or unexpectedly drop a connection.

What can they do now? Not much, one must admit. And yet I sometimes think about phone booths in light of these lines from Emily Dickinson's poem "Four trees upon a solitary acre":

> Four Trees—upon a solitary Acre—
> Without Design
> Or Order, or Apparent Action—
> Maintain[3]

Here is a cluster of objects with no ascribable function, action, or organizing principle that nevertheless "maintain," a verb suggesting preservation and continuance, as well as support and provision. It is not clear who or what these trees maintain; perhaps only themselves, perhaps the world. The image is stark, but also dignified.

The last time I received a call from a pay phone, it was from an old friend who had lost his wife, his job, his house, and was living in his car. He parked in the lot of a local store and used a nearby pay phone before the store closed he moved on to another part of the city. Though he'd had a cell phone at one point, he had temporarily lost that, too. Weeks would pass without my hearing from him, and then I might receive two or three calls in a row, nothing urgent, he would emphasize, just calling to check in. In spite of all appearances to the contrary he was always "wonderful, just wonderful." For a while I spent a lot of time by the side of a reservoir recording his life story, because his life seemed far more interesting than mine, and he was a good storyteller, and it was something to do, for both of us. It took us a long time, maybe thirty hours of talking, to circle back to the present, by which time we had gone through his entire childhood in the south, his mother's several marriages and divorces, his own several marriages and divorces, books, estates, children. Without any of the accouterments one is supposed to have as an adult, he felt like "one of the invisibles." Without a job, without a house, he'd been reduced to something more akin to landscape, mood—inflecting the city with meaning but unrecognized by the city itself.

Now the café is gone, replaced with a convenience store, and so is the friend, gone to another city. The pay phone no longer has a dial tone, and everything else about it seems muzzled, with ads, grit, even sweat—as if the perspiration of multiple lives lived at some personal cost has coalesced into something solid and smothering. Ads for taxies plaster its surfaces. The obligatory graffiti inscribes the sides of the enclosure with enigmatic hieroglyphs that must have been left some time ago, when a clean surface might have seemed egregious and ripe for being defiled. In a wealthier area perhaps it would have been removed already but here it's been overlooked or ignored and is now a repository for dust, trash, and memory.

FIGURE 22 Phone kiosk, taken by David Eng.

"I wouldn't say there was anything very mysterious about that phone booth," remarked Dennis Casebier, a historian of the Mojave area, when I asked him what he knew about the Lanfair booth.[4] My own attempt to find the Lanfair and Mojave phone booths had been an utter failure. Relying on the scant information provided by the Internet, my husband and I ended up eighteen miles up an unpaved road in the Mojave National Preserve about fifteen miles south of where we needed to be. We only learned that after pulling over and talking to a pair of hunters, one of whom at least seemed to know the Lanfair we were talking about, but not the phone. It was beautiful country, though—big sky, Joshua trees, miles of sage—and I could see why people might have been mesmerized by the idea of a piece of technology, even an obsolescent one, existing this far out. At one point we crossed paths with a Union Pacific train we had been tracking since we had been thirty minutes outside of Vegas, on the 1-5. In the distance sunlight shimmered off a series of solar panels steadfastly powering a small corner of the grid.

I thought of a passage from Watson's autobiography, in which he describes traveling to North Conway, New Hampshire to experiment with installing a long distance line between North Conway and Boston. The experiment yielded disappointing results but what stood out for Watson were the mountains: "Those I saw on that winter day were to me like a glimpse into heaven, or, into some mystic other-world I had dreamed of. . . . Up to that time the telephone had filled my life completely, but after that day in North Conway the telephone shrank a little."[5] Watson left the phone business in 1880, when "the income question was settled," ultimately establishing a shipyard that would become one of the major suppliers during the Second World War.

"Here is unfenced existence," writes Philip Larkin, "Facing the sun, untalkative, out of reach."[6] Even in the middle of the Mojave, though, I didn't feel out of reach, nor was I. Whether we are aware of it or not we are constantly involved in complex triangulations of data, we are always being monitored, targeted specifically or identified as part of group whose actions are being distilled into patterns. Every day there are

FIGURE 23 Phone kiosk. Photo © David Eng, used with permission.

numerous attempts being made in every media possible to understand how this networked world is changing our minds and our identities. It's the simple truth, though, that by the time anyone understands what's happening to us—how we are being transformed by the dizzying pace of technological change—we will be long gone.

Now I walk through the streets of my city in the way Stephen Dedalus once walked through the streets of Dublin, explaining to his friend Cranly that although "only an item in the catalog of Dublin's street furniture," a clock on the side of the Ballast Office, is "capable of epiphany."[7] What's happening, says Dedalus, is the spiritual eye finding its point of focus. Here, in Los Angeles, I catch fragments of conversations that are in conversation with each other: "It was a regrettable summer. . . ." "We had to be careful. . . ." "These parades look tired. . . ." Perhaps these aren't fragments of conversations I've heard but fragments of poems. Perhaps I am mishearing everything. Perhaps I feel more natural this way, looking aimlessly for shelter in a world that traverses me so easily.

NOTES

Chapter 1

1 Masataka Yamaura, "Overturned phone booths used to 'contact' tsunami victims to be repaired," *The Asahi Shimbum*, January 9, 2015.

Chapter 2

1 Gaston Bachelard, *The Poetics of Space* (Boston: Beacon Press, 1994), 31.

2 James Merrill, as quoted by Rachel Hadas in *The American Scholar*, accessed January 15, 2015, https://theamericanscholar.org/to-feel-and-keep-the-eyes-open/#.VOyWVktstuY.

Chapter 3

1 *Trading Places*, directed by John Landis (1983; Los Angeles, Paramount Pictures).

2 Joan Didion, *Play It As It Lays* (New York: Farrar, Straus and Giroux, 1970), 101.

3 Jimmy Stamp, "The Pay Phone's Journey From Patent to Urban Relic," *Smithsonian.com*, September 18, 2014, accessed January 23, 2015, http://www.smithsonianmag.com/ist/?next=/history/first-and-last-pay-phone-180952727/.

4 Anonymous, "Times Sq. Crowd Roars For Both," *The New York Times*, July 3, 1921.

5 Marcel Proust, *Swann's Way*, trans. Montcrieff and Kilmartin (New York: The Modern Library, 1992), 59–60.

6 Dictionary.com, accessed January 15, 2015, http://dictionary.reference.com/browse/angle+of+repose.

Chapter 4

1 Brenda Hillman, "Phone Booth," in *Practical Water* (Middletown, CT: Wesleyan University Press, 2011), 17.

2 Thomas Pynchon, *Bleeding Edge* (New York: The Penguin Press, 2013), 244–45.

Chapter 5

1 Gay Talese, "Phone Booths to Blossom Out in Glass and Gold Aluminum," *The New York Times*, October 20, 1961.

2 Anonymous, "The Solid Gold Telephone Booth," *The New York Times*, October 25, 1961.

3 Brochure, American Telephone Booth Company, October 20, 1882.

4 Thomas A. Watson, *Exploring Life* (New York: D. Appleton and Company, 1926), 116.

5 Watson, *Exploring*, 83.

6 Ibid., 81.

7 Alec Wilkinson, "A Voice From The Past," *The New Yorker*, May 19, 2014.

8 A. J. Liebling, *The Telephone Booth Indian* (New York: Broadway Books, 2004), 37.

9 Anonymous, "Orders Phone Booths Out," *The New York Times*, February 16, 1929.

10 Cullen Bryant Colton, "The Booth," *Western Electric*, October 1949.

11 Andy Beckett, "Remember the Phone Box," *The Guardian*, November 10, 2002.

12 John Sutherland, "When phone boxes come into their own," *The Guardian*, July 10, 2005.

13 Adam Tompkins, private correspondence with Ariana Kelly, February 12, 2015.

14 Beckett, "Remember the Phone Box."

15 Anne Carson, "Kinds of Water," in John D'Agata (ed.), *The Next American Essay* (Minneapolis: Graywolf, 2003), 192.

16 Marcel Proust, *The Guermantes Way* (New York: Modern Library, 1998), 175–76.

17 Proust, *Guermantes*, 176.

18 William Caughlan, private correspondence with Ariana Kelly, December 3, 2014.

19 "Those Handy Public Telephones," *Pacific Telephone Magazine* (June 1958): 0–2.

20 Cullen Bryant Colton, "The Booth," *Western Electric*, October 1949.

21 Bell Telephone System, advertisement.

Chapter 6

1 Mark Twain, "Christmas Greetings," *The Boston Globe*, December 25, 1890.

2 Timothy Wu, *The Master Switch* (New York: Alfred A. Knopf, 2010), 25.

3 "Surprising, But Not Inexplicable," *The New York Times,* August 19, 1905.

4 Eula Biss, "Time and Distance Overcome," in *Notes from No Man's Land* (Saint Paul, MN: Graywolf, 2009), 3–6.

5 Claude S. Fisher, *America Calling* (Berkeley: University of California Press, 1992), 66–67.

6 James Thurber, *Lanterns and Lances* (New York: Harper & Brothers, 1961), 20.

7 Robert Richards, "The Big E Comes Home—Still Queen," *United Press International*, October 17, 1945.

8 "Pier Phone Booths Rushed By Sailors," *The New York Times*, October 18, 1945.

9 Lizette Alvaraz, "An Internet Lifeline for Troops in Iraq and Loved Ones at Home," *The New York Times*, July 8, 2006.

10 Pew Research Center, accessed February 21, 2015, http://www.pewinternet.org/2013/06/05/smartphone-ownership-2013/.

11 Jean Baudrillard, *The Ecstasy of Communication*, accessed December 1, 2014, http://iris.nyit.edu/~rcody/Thesis/Readings/The%20Ecstacy%20of%20Communication%20-%20Baudrillard.pdf.

12 Avital Ronell, *The Telephone Book: Technology, Schizophrenia, Electric Speech* (Lincoln, NE: University of Nebraska Press, 1989), 5.

13 Dan Morse, "Still Called by Faith to the Phone Booth," *Washington Post*, September 3. 2006.

14 Ronell, *Telephone* Book, 3.

15 Mircea Eliade, *The Sacred and The Profane: The Nature of Religion* (New York: Harcourt Brace & Company, 1987), 49.

16 Georg Simmel, "The Metropolis and Mental Life," *The Blackwell City Reader* (Oxford, UK and Malden, MA: Wiley, Blackwell, 2002), accessed December 15, 2014, http://www.esperdy.net/wp-content/uploads/2009/09/Simmel_21.pdf.

Chapter 7

1 Sierra Nelson, private correspondence with Ariana Kelly, November 21, 2014.

2 Jason Brown, private correspondence with Ariana Kelly, November 21, 2014.

3 Walter Benjamin, "Berlin Childhood around 1900," in Marcus Paul Bullock and Michael William Jennings (eds), *Selected Writings, Volume III, 1935-1938* (Cambridge, MA: Harvard University Press, 2003), 349.

4 *The Matrix*, directed by the Wachowskis (1989; Burbank, CA: Warner Bros. and Village Roadside Pictures).

5 Matrix Resolutions, accessed February 15, 2015, http://www. matrixresolutions.com/index.html?page=saq_hardlines.

6 *The Matrix*.

7 Mark Thomas, *The Pay Phone Project*, http://www.payphone-project.com.

8 Jill Lepore, "The Man in the Box," *The New Yorker*, November 11, 2013.

9 Virginia Woolf, *To the Lighthouse* (New York: Harcourt Brace & Company, 1981), 138.

10 As quoted by "Hello, Pay Phone Information? Enthusiast Provides the Answer," *The New York Times*, May 13, 2004.

11 As quoted by Tanvi Misra, "Why Some Places Still Have Plenty of Pay Phones," *The Atlantic's CityLab*, November 10, 2014, accessed January 15, 2015, http://www.citylab.com/tech/2014/11/why-some-places-still-have-plenty-of-pay-phones/382454/.

Chapter 8

1 *Red*, directed by Krzysztof Kieslowski (1994; Santa Monica, CA: Miramax Films).

2 Nora Kelly and Bill Jeffries, private correspondence with Ariana Kelly, November 21, 2014.

3 Roland Barthes, *A Lover's Discourse* (New York: Farrar, Straus and Giroux, 1978), 37.

4 Melanie Noel, private correspondence to Ariana Kelly, January 13, 2015.

5 "The Telephone at the Paris Opera," *Scientific American*, December 31, 1881, accessed December 1, 2014, 422–23, http://earlyradiohistory.us/1881opr.htm.

6 Sierra Nelson, private correspondence to Ariana Kelly, November 21, 2014.

7 Ben Wurgaft, private correspondence to Ariana Kelly, November 22, 2014.

8 Samuel Warren and Louis Brandeis, "The Right to Privacy," *Harvard Law Review* 4 (1890), accessed December 1, 2014, http://faculty.uml.edu/sgallagher/Brandeisprivacy.htm.

9 "What Every Woman Wants to Know," *The New York Times*, January 10, 1915.

10 Timothy Wu, *The Master Switch* (New York: Alfred A. Knopf, 2010), 101–14.

11 Jessica Gelt, "Beverly Hills Hotel marks 100 years as stars' secret retreat," *Los Angeles Times,* May 22, 2012.

12 Erik Davis, *Techgnosis* (New York: Harmony Books, 2004), 82.

13 Justice Louis D. Brandeis, "Olmstead v. United States: The Constitutional Challenges of Prohibition Enforcement," *Olmstead v. United States*, 277 U.S. 438 (1928), accessed December 1, 2014, http://www.fjc.gov/history/home.nsf/page/tu_olmstead_doc_15.html.

14 *Katz v. United States*, 389 U.S. 347 (1967), accessed November 1, 2014, http://www.uscourts.gov/educational-resources/get-involved/constitution-activities/fourth-amendment/wiretaps-cell-phone-surveillance/facts-case-summary.aspx.

15 Renata Adler, *Speedboat* (New York: NYRB Classics, 2013), 100.

16 Francis X. Clines, "Massena Enjoys Seaway Prosperity After Long Wait," *The New York Times*, June 27, 1969.

17 "Hundreds of Devices Hidden Inside New York City Phone Booths," BuzzFeed, October 6, 2014.

Chapter 9

1 Albert Camus, *The Myth of Sisyphus and Other Essays* (New York: Vintage International, 1991), 15.

2 Huruki Murakami, *Sputnik Sweetheart* (New York: Vintage International, 2002), 8.

3 Murakami, *Sputnik*, 135.

4 Ibid., 208–09.

5 Ibid., 84.

6 Ibid., 209.

7 David Trotter, "The Person in the Phone Booth," *The London Review of Books* 32, no. 2 (November 28, 2010): 20–22.

8 F. Scott Fitzgerald, *The Great Gatsby* (New York: Scribner, 1925), 2.

9 *The Fly*, directed by David Cronenberg (1986; Century City, CA: 20th Century Fox Film Corporation).

10 Larry Tye, *Superman* (New York: Random House, 2012), 96.

11 Steven Kurutz, "Urban Tactics; Sorry, Superman," *The New York Times*, December 8, 2002.

12 Joseph Brodsky, "In a Room and a Half," *The New York Review of Books*, February 27, 1986.

Chapter 10

1 John M. Glionna, "Reaching Way Out," *Los Angeles Times*, September 18, 1999.

2 Glionna, "Reaching Way Out."

3 Joe Rosenberg, (2014) "Mojave Phone Booth," *Snap Judgment*, accessed December 7, 2014, http://snapjudgment.org/mojave-phone-booth.

4 *Snap Judgment*.

5 *Mojave Mirage*, directed by Derek Roberto and Kaarina Cleverley Roberto (2003; documentary).

6 Davis, *Techgnosis*, 208.

7 Brad Templeton, "Free phone booth at Burning Man," 2014, accessed November 15, 2014, http://www.templetons.com/pq/.

8 David Lazar, "Calling for His Past," in *Occasional Desire* (Lincoln, NE: University of Nebraska Press, 2013), 6–7.

9 Chloe Taylor, *The Culture of Confession from Augustine to Foucault* (New York: Routledge, 2009), 47.

10 Robert Bernasconi, "The Infinite Task of Confession: A Contribution to the History of Ethics," *Acta Institutionis Philosophiae et Aestheticae* 6 (1988): 80.

11 Michel Foucault, *The History of Sexuality*, trans. Robert Hurley (New York: Pantheon Books, 1978) 61–62.

12 Eliade, *Sacred*, 20.

13 Ibid., 21.

14 Ibid., 25–26.

15 Dennis Casebier, private correspondence with Ariana Kelly.

Chapter 11

1 APCC, accessed February 21, 2015, http://www.apcc.net/i4a/pages/index.cfm?pageid=40.

2 Junichiro Tanizaki, "In Praise of Shadows," in Phillip Lopate (ed.), *The Art of the Personal Essay* (New York: Anchor Books, 1995), 339.

3 Emily Dickinson, "742," in Thomas H. Johnson (ed.), *The Complete Poems of Emily Dickinson* (New York: Little Brown and Company, 1961), 364.

4 Dennis Casebier, private correspondence with Ariana Kelly, December 14, 2014.

5 Watson, *Exploring,* 104.

6 Phillip Larkin, "Here," in Anthony Thwaite (ed.), *Collected Poems* (New York: Farrar, Straus and Giroux, 1993), 136.

7 James Joyce, *Stephen Hero* (New York: New Directions, 1959), 211.

ACKNOWLEDGMENTS

Many thanks to Betalevel's Errata Salon, which gave me the impetus to write the essay that inspired this book, and to Evan Kindley, whose edits made the original essay much more substantive and coherent. Thanks to Christopher Schaberg, Ian Bogost, and Haaris Naqvi for their editorial and production support. Thanks to David, Ian, June, and Nora Kelly, as well as Bill Jeffries, Amber Caron, Zachary Greenberg, Adam Rose, and Ryan Wilson for looking at multiple drafts of various chapters. Thanks again to Nora Kelly, who not only offered invaluable suggestions about the book's content and structure, but also copyedited the entire manuscript. Thanks to all of the people who contributed anecdotes, references, and photos, many of which ended up in the book. Thanks to David Lawrence Eng and Lia Eng for the coffee that fueled most of this project. Finally, infinite love and gratitude for David Gene Eng, who gathered all of the images in this book, the best of which are his own, and who was a constant source of love, encouragement, and laughter.

INDEX

Page references for illustrations appear in *italics*.